SEMINAR STUDIES IN HISTORY

Editor: Patrick Richardson

COAL MINING IN THE EIGHTEENTH AND NINETEENTH CENTURIES

SEMINAR STUDIES IN HISTORY

Editor: Patrick Richardson

A full list of titles in this
series will be found on the
back cover of this book

SEMINAR STUDIES IN HISTORY

COAL MINING IN THE EIGHTEENTH AND NINETEENTH CENTURIES

Brian Lewis

Lecturer in History
Scawsby College of Education
Doncaster

LONGMAN

LONGMAN GROUP LIMITED
London

ASSOCIATED COMPANIES, BRANCHES AND
REPRESENTATIVES THROUGHOUT THE WORLD

© Longman Group Ltd 1971

All rights reserved. No part of this publication
may be reproduced, stored in a retrieval system,
or transmitted in any form or by any means,
electronic, mechanical, photocopying,
recording, or otherwise, without the prior
permission of the Copyright owner.

First published 1971

ISBN 0 582 31422 4

PRINTED IN GREAT BRITAIN BY
WESTERN PRINTING SERVICES LTD, BRISTOL

Contents

INTRODUCTION TO THE SERIES	vii
ACKNOWLEDGEMENTS	viii
FOREWORD	ix

Part One · The Background

1 COAL MINING BEFORE 1700	3

Part Two · The Industry

2 THE GROWTH OF THE MINING INDUSTRY	15
3 MINING COMMUNITIES	28
Social Conditions	28
Working Conditions	41
4 COAL MINING LAW	54
5 THE COAL MINERS' TRADE UNIONS	66

Part Three · The Consequences

6 THE AFTERMATH	87

Part Four · Documents

DOCUMENTS	93
BIBLIOGRAPHY	118
INDEX	125

Introduction to the Series

The seminar method of teaching is being used increasingly in VI forms and at universities. It is a way of learning in smaller groups through discussion, designed both to get away from and to supplement the basic lecture techniques. To be successful, the members of a seminar must be informed, or else—in the unkind phrase of a cynic, it can be a 'pooling of ignorance'. The chapter in the textbook of English or European history by its nature cannot provide material in this depth, but at the same time the full academic work may be too long and perhaps too advanced for students at this level.

For this reason we have invited practising teachers in universities, schools and colleges of further education to contribute short studies on specialised aspects of British and European history with these special needs and pupils of this age in mind. For this series the authors have been asked to provide, in addition to their basic analysis, a full selection of documentary material of all kinds and an up-to-date and comprehensive bibliography. Both these sections are referred to in the text, but it is hoped that they will prove to be valuable teaching and learning aids in themselves.

Note on the System of References:

A bold number in round brackets (**5**) in the text refers the reader to the corresponding entry in the Bibliography section at the end of the book.

A bold number in square brackets, preceded by 'doc.' [**docs 6, 8**] refers the reader to the corresponding items in the section of Documents, which follows the main text.

PATRICK RICHARDSON
General Editor

Acknowledgements

The author wishes to thank the following for their assistance in the preparation of this work: John Goodchild, Curator Cusworth Hall Museum, Doncaster, The North of England Institute of Mining and Mechanical Engineers, The National Coal Board, George Purdom, John Martlew, and Sheila Oldfield.

Foreword

In the Midlands coalfield in the nineteenth century the signal which ended a shift was the cry 'lillycock' but in the North-East the same message was conveyed by the shout 'kenner' or 'loose all'. An underground chargehand was a 'doggy' in South Staffordshire but a 'corporal' in Yorkshire; Dudley 'kibbles' were North Welsh 'cranks' and these to the uninitiated were small coals. Immediately one starts to study coal-mining history it is not only apparent that a glossary of terms is needed but also that regional differences make generalisation extremely dangerous. Witnesses before the Select Committees on coal in the last century constantly reiterated the viewpoint that it was dangerous to bring in general safety legislation because the nature of pits and coal mined varied enormously. Local differences are considerable and it is essential to realise from the outset that an essay which gives background to only a selection of documents will present a general rather than a full picture. It cannot look at all mining areas nor can it deal with every aspect of the industry. Space is limited and regrettably many important features of the industry's development, which would deserve careful consideration in a longer work, here have either fleeting reference or are ignored.

Part One

THE BACKGROUND

1 Coal Mining before 1700

The Romans were pragmatists and where coal came to the surface they used it. Unfortunately outcrop coal is poor fuel. Exposed to frost and rain it is subject to constant weathering, gives out little heat and produces noxious fumes. Only ease of access makes it in any way attractive and for this reason coal or its ash is usually found on Roman civilian sites within easy reach of known coalfields. Scraps of evidence exist which show that in some areas there was a coal trade which extended beyond the purely local market. Usually, though, coal was burned in the immediate vicinity of the mine. There is proof of traffic along the fenland rivers from the Nottinghamshire coalfield and microscopic analysis of coal found on the site of the York crematorium shows that it was mined over twenty miles away in the Leeds area. In both cases water transport was decisive.

Archaeologists have found coal traces on many Roman sites which suggests that many existing coalfields have a mining history of almost two thousand years. The Scottish field supplied the bunkers on the Antonine Wall and the Northumberland coalpits the guardhouse store at Housesheads. Lancashire mines provided coals for metal workers at Wilderspool, near Warrington, and Yorkshire supplied some of the fuel which fired kilns at Cantley, Doncaster. Central-heating hypocausts at Wroxeter burned Shropshire coal; those at Caerwent took supplies from South Wales. There is evidence for the use of coal in the Bristol area in an account of its presence on the altar of the temple of Sulis Minerva in third-century Bath. But although this attests widespread usage for varied purposes it does not override the fundamental principle that coal was used only when easily available. In general wood or charcoal was preferred.

Following the Roman occupation, evidence of mining is slight and questionable. A reference to twelve hundredweight of coal being given to Peterborough Cathedral in 852 is thought to be a misinterpretation and there is no mention of mining in the Domesday Survey in 1086. There are almost a thousand blank years until we come to

The Background

recognise the origins of the industry in that quickening of the tempo of economic life which occurs in the thirteenth century (**1**).

Frequent references to coal mining appear at that time but it should not be thought that this necessarily reflects greatly increased demand. It is rather that evidence is more readily available; it exists in the forms of State papers, manor court rolls and ecclesiastical records—and in the personal documents, such as wills, inventories and contract agreements (**3**). There is a charter of 1239 from a coroner's report from Birstall, Yorkshire (1287), on a man killed in a pit. Written evidence of monastic interest in mines is widespread and proof of mining in the Forest of Dean is given by a summons requiring the Free Miners to be present at the siege of Berwick in 1310 where it is presumed they were needed to dig saps. Yet despite these and numerous other cases our knowledge is still fragmentary and our conclusions tentative. A list of dates and a few isolated facts prove little. We can make guesses and surmise that mines were small and inland trade, lacking developed transport facilities, localised, but with present evidence it is difficult to argue that output rose. Resting on insufficient written proof the expansion may be apparent rather than real.

If the thirteenth century saw the slow emergence of an industry it was the fourteenth century that saw its rapid growth. Although most of the fields we know today were by that time producing coal, expansion was uneven. Inland fields continued to support family units. These worked outcrop mines or dug elementary open-cast workings but in the area surrounding Newcastle-upon-Tyne a thriving and complex industry appeared (**4**).

Monopolists, shipping magnates and recognisable coal capitalists emerged in the reigns of Edward III and Richard II and it is not difficult to see why this should have happened. The Tyne and North Sea made it possible to send coal to London, the populous south-east and the Continent thereby establishing pre-eminence over the inland fields which did not enjoy such transport and market facilities. Wily speculators appeared. When a fifteenth-century parliament levied a twopenny tax on each chaldron of coal sold to non-freemen of Newcastle and based collection on the assumption that a London-bound coal keel carried twenty chaldrons the merchants evaded full payment by building boats which would carry twenty-three chaldrons. An Act of 1432, which caused the load line of boats to be marked with nails and threatened forfeiture on deliberate tax

Coal Mining before 1700

evasion, attempted to stop this practice. This parliamentary interest is an indication of the magnitude of the Newcastle trade for legislation for tax assessment purposes is in itself recognition of economic well-being. Although small when compared with the 800,000 tons of coal produced in the area at the end of the seventeenth century, 70,000 tons was exported in 1377 and this figure was not surpassed a century and a half later. The Newcastle fleet which is known to have carried coal to London in 1270 left its mark on the capital in street names. The clean-air legislation proclaimed in 1307 because of the menace of sea-coal fires together with the fifteenth-century introduction into general use of chimney-pots are a legacy of medieval colliery expansion in the north-east.

The story of this regional expansion has been so frequently told that it is possible to neglect developments in lesser coalfields, yet investigation of other areas can be profitable for in the small fields of Yorkshire and Warwickshire the problems which beset eighteenth- and nineteenth-century miners are seen in their primitive beginnings. Let us take a look, for example, at the Earl of Shrewsbury's small pit in Sheffield Park (5). Here the employees never exceeded eight and in 1580 it produced 1,316 tons of coal which was sold in the immediate locality. The chief hazard which hampered mining operations was 'damp' and in a period of a possible 156 working days work took place on only fifty-three recorded occasions because of the presence of this gas. When a second shaft was sunk ventilation improved and there was increased productivity. Water was another hindrance which restricted production.

The colliers worked in a team which was led by the banksman and included a pykeman (a hewer) and a barrower. Holidays were regular and absenteeism common. 'Monday after the fair day, they wrought not' has a modern ring, as do the unexplained holidays before St James's Day. At Christmas they took a week's holiday, at Easter four or five days and at Whitsun three. Thirteen saints' days in addition to Candlemas, May Day, Midsummer Day, Haw Thursday and Cocking Monday were also celebrated. The frequency of these holidays is a legacy from the medieval period which the miner's conservatism made him reluctant to relinquish. A critic of South Wales miners at this time stated that they 'observe all abolished holy days and cannot be weened from that folly' (7). This observation could have been made of the Sheffield men and yet it is possible to recognise sound economic motives in the abundance

The Background

of official and unofficial holidays. Fear of over-production and involuntary short time, as well as the demand for increased leisure time, can cause miners to limit and thus control production. The call 'Eight hours' work and eight hours' play, eight hours' sleep for eight bob a day' which united Yorkshire miners in the 1880s was motivated by the same principles. Absenteeism, over-production, ventilation problems, water hazards, complicated labour structures, conservative attitudes and group payment systems are all met within a study of this small Sheffield pit, and these are the principle themes which emerge in coal-mining history in the eighteenth and nineteenth centuries.

Many early mines were drift mines. Where coal outcropped, miners drove galleries into the hill or river bank and worked them until driven out by water, gas or the fear of roof falls. Elementary drainage was achieved either by driving an adit or 'sough' from a lower level until it met the floor of the working or by employing water drawers with buckets. The former method was more efficient but was also dangerous. At Sheffield the specialist miner who broke through from the 'sough' to the mine was paid £2 for that operation at a time when water drawers received 3*d* to 6*d* a day. Drowning as the water flooded out was a real possibility; suffocation was another. Choke damp, sometimes called black damp, or styth, gathered in the shallow mines and threatened life. In the winter when the temperature in the workings was lower than outside some air circulation did occur, but when in the summer months the reverse happened explosive, suffocating mixtures collected in the mine. They could, unlike fire-damp, be driven out by wafting but often pits were unproductive because of the presence of gas. Book entries such as 'three days lost by reason of damp' occur in weekly sixteenth-century account records. As the mines got deeper the problem increased.

Shaft coal mines, thirty feet in depth, are known from the fifteenth century but the history of this type of working is longer. The Neolithic flint mines known as Grimes Graves on the Norfolk and Suffolk border are shaft and gallery mines. When the shaft does not extend into galleries but bells out at the bottom so that its cross-section resembles an inverted cone it is called a 'bell pit'. Used as early as the thirteenth century batteries of these mines pockmarked the landscape, their entrance shafts often no more than ten yards apart. In some areas they became subject to elementary industrial

Coal Mining before 1700

agreement, as when the Free Miners of the Forest of Dean (**6**) ruled that shafts should not be sunk closer to each other than the distance a man could throw a shovel-load of rubbish from the pit he had worked out. Industrial archaeologists sometimes recognise the remains of these landscape features, but find it difficult to distinguish iron stone from coal bell pits.

Descent into the shaft pit was usually by ladder and family units exploited a mine; the father was the hewer and the mother and children, using wicker baskets and barrows, acted as carriers. Under such circumstances mining was not always the coal worker's sole occupation, but only provided part-time employment, the rest of his time being spent working on the land. Pitmen, as late as the seventeenth century, are sometimes recorded as 'husbandmen', but by the end of the Middle Ages, and especially in the Tudor period (**7**), changes necessitating specialisation and the employment of more men were taking place in the north-eastern coalfield.

Although small pits continued to be viable in the Tudor period intensive production mines of considerable investment appeared in the north-east. Queen Elizabeth I entered the coal trade when in 1582 she leased from the Bishop of Durham the manors of Whickham and Gateshead. The assurance of the payment of a sum of one shilling per chaldron on all exported coal caused her to grant the Society of Hoastmen in Newcastle the sole right to load and unload coal (**8**). This meant that twenty men could manage the output or vend coal as they liked, also regulating its price. Their monopoly made the price of coal quickly rise from four shillings to nine shillings per chaldron.

Yet, despite monopoly and inflation, coal was in great demand and production increased. As the shallow pits on Tyneside became worked out owners were faced with the alternative of digging deeper or fanning out laterally, seeking coal farther inland. Neither was easy and both were attempted. In the sixteenth century they tended to choose the former because of the difficulties of overland transport of a bulk commodity, but these deeper pits brought new problems. Search boring and shaft sinking were expensive. Depth intensified the water problem, necessitating the employment of water-raising machinery, and gaseous 'fire-damp' was encountered. Since technical improvements did not offset these difficulties, output per man did not advance and increased production was only achieved when greater numbers were employed. In 1551 the annual shipment of

The Background

coal from Durham and Northumberland was 36,000 tons of which 97·2 per cent came from the Tyne and only 1·4 per cent from the Wear Valley. By 1631 it was 522,000 tons and in 1690 800,000 tons, of which 73·5 per cent came from the Tyne Valley, 22·5 per cent from the Wear Valley and 4 per cent from the Northumberland coast. This dramatic increase and the tendency to expand into hitherto inaccessible areas because of the development of tramways and other transport features is noted in other coalfields.

The overwhelming demand for coal which at the beginning of the seventeenth century caused this rapid expansion was partly due to society's inability to maintain traditional fuel supplies. The old adage that in the sixteenth century sheep ate up men, that animal husbandry encroached on arable land and reduced total cultivated acreage needs qualification. Although it is true that food-producing fields were used for sheep rearing, the chain reaction which followed must not be ignored. In an attempt to maintain grain supplies the farmer in his turn attacked woodland and waste. Timber supplies were therefore cut as woods became ploughland, and this at the very time when the demand for timber was increasing (9).

If the eighteenth and nineteenth centuries were the age of iron and steel, then the sixteenth was the age of wood. There were wooden houses, wood fires, timber ships and many of the domestic articles, buckets, spoons, plates, which are now available in coloured plastic, were then made of wood. If fashions changed the forest suffered. Brick houses replaced Tudor 'black and white', but more timber fuel was needed to fire the brick than would have been required to build a wooden house. Eight stout beech trees rendered into charcoal would only convert ore into one ton of bar iron. When in the seventeenth century metal production exceeded 35,000 tons a year the smelters required 200,000 stout trees. Glass manufacturing was even more wasteful, and glassmakers cleaved a line through the woods of the Severn Valley as they moved in search of timber. It is reported that in his lifetime one Durham man felled 30,000 trees. Deforestation on this scale could not easily be checked. Legislation in Queen Elizabeth's reign sought to halt it but could not. The arcadian longing of the Shakespearean Duke in his Arden is a nostalgic desire for a landscape which the Elizabethans saw shrinking and almost passing away. In the end it was the prohibitive cost of wood which caused man to look for an alternative fuel. The exploitation of coal resources was an answer. Industries which had for

Coal Mining before 1700

centuries burned wood or charcoal began to burn coal and coal production advanced.

In some industries the substitution of coal for wood presented no difficulty, but in others its use had disadvantages. Where it was the traditional fuel it continued to be used but in greater quantity. Lime burners used it and lime was present in mortar. The increased amount of brick and stone construction which occurs in the seventeenth century therefore bears directly on increased coal production. The improved husbandry which accompanies the realisation and exploitation of lime as a fertiliser can also be linked to colliery expansion.

Other industries came to use this fuel as wood supplies either ran out or became an uneconomical proposition. By 1700 coal had replaced wood as the main fuel in salt making in most of the major centres of that industry. Gunpowder depended on saltpetre, and saltpetre in its turn learned to depend on coal. Sugar, soap, starch, candle, dye and textile manufacturers, meat-bakers and brewers, all needed coal and burned it in great quantity.

However, there were other industrial processes in which coal substitution was not easy. When hops and malt were dried with coal the beer had a sulphurous taste, nor could coal be used to produce quality iron. Cellini's account of casting 'Perseus', when the household pewter was thrown in to overcome curdling, shows smelting and casting to be delicate arts. As with bronze so with iron, for to produce this metal from its ore high temperatures and complex chemical changes are needed. Carbon has to replace the ore oxides and unite with the metal. Charcoal, because of its relative purity, can accomplish this but most coals cannot. Here the presence of extraneous elements, such as sulphur, inhibit the process and only allow a crude metal to be founded. The problem of the ironmaster, therefore, was either to free the coal of its damaging properties or to invent a device to protect the raw materials from the flames and fumes. He chose the former, baked the coal and produced a crude form of coke.

There is evidence of this type of carbonisation by the end of the sixteenth century. Sir Francis Willoughby of Wollaton, Nottingham, in a letter to the Earl of Rutland, dated 1586, mentions a man who made coke and a patent for a beehive coke oven granted to Sir William St John in 1620 further confirms its use. Dr Plot, the Staffordshire historian, commented on its general acceptance when in 1686 he wrote: 'they have a way of charring it [the coal] in all

The Background

particulars as they do wood, whence the coal is freed from these noxious steams that would give malt an ill odour. The coal thus prepared they call coke which conceives a heat almost as strong as charcoal itself, and is as fit for most uses, but for melting, fineing and refining of iron, which it cannot be brought to do, though attempted by the most skilful and curious artists'(11). From this account it is obvious that coking was efficient and widespread, but still ineffective when used to reduce ironstone. Until it could be made so, coal production was limited by a demand which did not include its largest potential customer, the ironmaster.

Still, by the middle of the seventeenth century, British mines were producing five times as much as the rest of the world. Mines became deeper and extended further underground. With the cry 'Treason' the ever-sensitive James I, having descended into a mine on the mainland, emerged from a shaft onto an insulated wharf in the centre of the Forth. A Newcastle coal merchant in 1649 was employing over 1,000 men, and colliery disasters in which nineteen, thirty, and sixty mine workpeople died are known. The majority of mines continued to be small, but there were large collieries in Scotland, Cumberland and the North-East.

As the pits grew bigger they became deeper. Deep mines brought new problems (10). The gas found farther underground could not be wafted away. Called sometimes 'sulphurous vapour' or 'surfet', it is more generally known as 'fire-damp'. Formed during the decay of vegetable matter and held under pressure between layers of coal, mining operations allowed this gas to escape. Liberated into the galleries, it fouled air currents and violently exploded on coming into contact with naked flames. Drainage could no longer be solved by driving fresh adits or by simply employing more bailers; ingenious machines were needed and invented. Bucket gins were used in Scotland and the North of England in the sixteenth century. Endless long linked chains, to which were attached wooden or leather containers, passed round a drum, and propelled by a water wheel descended the shaft, took water from the drainage sump, before surfacing and discharging at ground level. The Chinese chain pump, seen and described in 1672, was another method of raising water. It consisted of a tube divided into two compartments in which a chain circulated. Pieces of wood were attached to the chain and these almost exactly fitted the tube. They pushed a volume of water before them as they advanced. Clever as such devices were, they never

Coal Mining before 1700

fully overcame drainage difficulties, and the production of an efficient pump was seen to be the way to prevent water hampering mining operations.

Some of the new hazards were removed by 'trial and error' solutions. Fire-damp could be dispersed if a man dressed in wet sacking crawled along the galleries, dug a shallow pit, lay in it, covered himself with boarding and then dragged along a lighted candle. What seems an act of heroic proportions was at least a pragmatic answer to an industrial problem, but it was also dangerous and expensive. Other solutions had to be found, and by the end of the seventeenth century applied science came to the aid of the colliery worker.

Science was losing its virginity. What had once been pure and scholastic, a subject which interested gentlemen versed in classical precedent, was hustled by technology into the modern age. Echoes from a strong medieval tradition persisted, for alchemy and astrology were still taken seriously. Edward Boyle—described in a pleasantly naïve epitaph as 'Father of Chemistry and Uncle of the Earl of Cork'—found it worthwhile to dispute with men whose authority rested on the unproved theories of Greek philosophers. Newton might devote time to fruitless alchemy, but essentially that science, which was pure, theoretical and therefore often irrelevant, gradually became neglected. New men observed, recorded and founded disciplines on mathematics and experiment. One of the early papers of the Royal Society in 1675 investigated mine vapours, and the pumps of Savery, Newcomen and Smeaton depended on earlier research (**13**). It was the prologue to an industrial revolution; the scientists were framing the questions which others were to answer.

Part Two

THE INDUSTRY

2 The Growth of the Mining Industry

George Stephenson once said that the Chancellor of England should give up the woolsack and sit instead on a sack of coals. When he said this he probably reflected popular opinion, for people at the beginning of the last century saw coal as the basic commodity of the economy. That Britain's growing industrial power rested on twin pillars—coal and iron—was not in doubt. If, however, this attitude was put to men in 1700, they would surely have found it remarkable, for to them coal was the fuel they put on the household fire, used in bakery and malting furnaces and which heated the salt pans of Cheshire. In their minds its industrial worth was limited, but the eighteenth century had only reached its ninth year when, in a remote part of Shropshire, Abraham Darby managed to solve a problem which was to alter the course of British history. In 1709 he discovered how to roast coal so that he could smelt iron (**17**).

There was nothing new about the problem which Darby solved. Throughout the seventeenth century ironmasters had tried to produce quality iron and thereby utilise local coal deposits. One of them, Dud Dudley (**15**), had even claimed in a treatise entitled 'Mettalum Martis' (1655) that he had managed to do this. Dudley was the bastard son of Edward Sutton, Earl of Dudley and foundry owner, and his mistress Elizabeth Tomlinson, 'a base born collier's daughter', and this nativity—the natural conjunction of iron with coal—suggests that if the outcome depended on the stars, then this was the man to work the wonder. Unfortunately the outcome, if Dudley's details are correct, depended on the small coals of the South Staffs 'thick seam', and since modern scientific methods have failed to produce a quality iron from this source, then his claim would seem to be as illegitimate as his birth, and deserve no more credence than that given to other seventeenth-century patentees of coal-fired furnaces.

Abraham Darby, however, did succeed where Dudley failed, and therefore he can justifiably be considered the founder of the modern iron industry. The circumstances, however, of the discovery are

The Industry

obscure. We know that in 1709, having moved to Coalbrookdale from the Bristol area, he started to buy quantities of coal in excess of that required for normal domestic purposes. His book-keeping records of the time survive, and these show regular fuel payments to a Much Wenlock master collier, one of which is for £20 worth of coal (**17**). It is also known that this first Darby had been apprenticed to a maltings builder, and, since technology often advances in an erratic and unpredictable fashion, it might be assumed that he learned about the techniques of coking not from the patents of unsuccessful ironmasters but from makers of good quality beer. What is certain, however, is that Darby's invention was slow in spreading, and fifty years later knowledge of his invention was still confined to Shropshire and Denbighshire, but by the second half of the century more ironmasters, having knowledge of his techniques, began to buy coal in ever-increasing quantities. From that time onwards the ironfounder gradually became the coal-merchant's principal customer, and in his turn the coal-owner turned with increasing regularity to the ironworker for the new tools which were required to increase production. Richard Reynolds cast his first iron tram rails in 1767, and these revolutionised colliery haulage. The Ironbridge of the third Darby, with its 70 ft cast iron ribs, provided precedent for the large castings later used in pit-head winding and pumping machinery. Such innovation, strengthened by the inventions of Cort and later Bessemer, made possible the rapid expansion of two primary industries. In 1717 an estimated 18,100 tons of pig iron was produced in England and Wales, and almost an equal tonnage of bar iron was imported from Russia and Sweden. Three years later 20,500 tons of pig iron and 14,900 tons of bar iron were produced by charcoal burners, compared with only 400 tons by coke smelting. Yet, almost seventy years later, the situation had undergone a dramatic transformation. Charcoal furnaces had lost their importance, coke smelting was paramount, and Britain had come to rely less and less on imported bar iron. In 1788, 76,000 tons of pig and bar iron were produced in coke furnaces, compared with a meagre 2,500 tons by other methods. In 1806, for instance, only eleven furnaces were using this fuel, compared with 162 using coke. Mineral fuel replaced timber, and coal output advanced in step with iron production. In 1700 the total annual coal tonnage was $2\frac{1}{2}$ million tons, but a hundred years later the 10 million tons figure had been passed. Of course, not all was absorbed by ironmasters. A population explosion

The Growth of the Mining Industry

and movement into towns meant that coal, in ever-increasing quantities, was needed by brickworkers, brewers, glaziers and other manufacturers. Townsmen, remote from scrub or woodland, burned coal rather than wood at the family hearth. For this reason production increased, and in the proximity of ironstone, coal and limestone lies the motive force for nineteenth-century industrial wealth and power. One of the advantages of cheap iron was that machinery which hitherto had been built of traditional materials, wood and brass, began to be made of ferrous metals. This is particularly true of pumping engines. When Thomas Newcomen (23) built his early mining pumps at the beginning of the eighteenth century, he was severely restricted by limitations of cylinder size and could not make an efficient cylinder beyond a certain diameter. Performance was thus impaired, or, if brass parts were used, the engine was expensive to produce and maintain. From 1722 onwards, however, he bought iron cylinders from Coalbrookdale and in time was able to make pumps of larger dimensions. The early pump at Heaton, Newcastle (1733), had a thirty-three-inch cylinder, but one made for the William Pit, Whitehaven, in 1810 had an internal cylinder diameter of eighty-two inches. James Watt, at the end of the century, built on Newcomen's foundations. Benefiting from Wilkinson's invention of a boring mill which cut accurately, and from the production of cheap iron-casting, he established a reputation which makes many regard him as the inventor of the steam engine. Yet, although in terms of world history Watt's engine is of greater importance, the engine which merits equal attention when we study the growth of the coal industry is that invented by Thomas Newcomen and usually called the 'common' engine. Samuel Smiles was correct when he called it clumsy and said that its action was 'accompanied by an extraordinary amount of wheezing, sighing, creaking and bumping', but this was the first successful mine pump, and its general use made colliery expansion possible. There is no doubt that Watt's (21) engine was more efficient, precise and was beautifully made, but Newcomen's pump had some advantages, and it was these which ensured its continued use in the coalfield, even after Boulton and Watt machines were available. To the colliery manager the simplicity of the earlier engine was one of its most attractive features. If a Watt, or the later Cornish, engine broke down repair often necessitated skilled engineering, but a village blacksmith could repair the 'common' engine. It could also burn low-grade coal

The Industry

whereas Watt's machine had refined tastes. For these reasons the early pump was popular with managers, since there were no fuel transport costs and coal was burned in order to win more coal. Its main disadvantage was excessive fuel consumption, for being without a separate condensing cylinder, the rapid exchange of temperature necessary for vaporisation and condensation required great quantities of coal. James Watt recognised this when in his patent of 1760 he wrote: 'My method of lessening the consumption of steam and consequently of fuel consists of the following principles.' The adoption of the separate condensing cylinder effectively reduced coal consumption and this ultimately led to Newcomen pumps being replaced by more sophisticated machines.

By 1835 the Newcomen engine was losing ground to various rivals, and although in the shallower pits of the Midlands and Scotland it continued in use, in the deeper northern mines engines, evolved from Watt's double acting engine, were generally adopted. With the Newcomen engine passing from general use the new deeper collieries lost an important link with coal mining's eighteenth-century past, for, although Newcomen's technical innovation primarily belongs to the history of mechanical engineering, coal could not have been won in progressive quantities had there been no efficient pumping engine.

When the coal-owner began to abandon the 'common' engine and adopt the Cornish Engines, it was because circumstances had changed and he needed something more than a pump. The increased demand for coal meant that mines had to work to full capacity. Gone were the days when leasees would insert output restrictions into mining contracts. Coal was needed in great quantities and it was needed fast as the demands of iron and other industries grew almost beyond recognition. Improved haulage was essential, and Watt's powerful rotative engine provided it. From 1784 when such a machine was installed at the Walker Colliery, Northumberland, such engines were used for winding and haulage. Increased demand not only meant that pits had to be driven deeper, but also that they had to lift coal faster and in greater quantity. Whereas in 1835 the maximum output of one pit consisted of about 300 tons in twelve hours, fifteen years later 600 tons and 800 tons were obtained for the same period of time. This was possible because high-pressure engines capable of developing 150 h.p. evolved from Cornish Engines.

The Growth of the Mining Industry

The Elsecar colliery engine, which stands virtually intact in a small Yorkshire colliery village just outside Barnsley, testifies to the importance of steam engines in the history of colliery development at the turn of the eighteenth century. Built in 1795 its survival *in situ*, when other similar machines have been shipped to American museums, makes it noteworthy, but its presence is of additional interest to the economic historian for it is in close proximity to eighteenth-century transport features (**25**).

The development of the Yorkshire coalfield, in which Elsecar stands, was uneven. It was served by three rivers, two of which, the Calder to the north and the Trent to the south, were navigable, and the Upper Don, which was not. For this reason in the early eighteenth century the coalfields of Wakefield and Sharlston and those of Chesterfield and Nottinghamshire developed, but the central area between Barnsley and Sheffield, being miles away from a navigable river, was stunted as its coal resources were only used domestically and by the iron and cutlery trades in Sheffield and Barnsley. This changed when waterway improvement schemes were implemented. By Acts of 1733 and 1751 the river Don was made navigable to beyond Rotherham. Coal was carried along its complete length and there was a rapid expansion of the mining industry in the area. By mid-century Yorkshire coal, much of it from this and the Sheffield area, was competing in the Humber estuary with Durham coal, and by 1769 it is estimated that 300,000 tons was produced annually. Canals further improved transport amenities, for the Dearne and Dove Canal (1793), which linked with the Barnsley Canal (from the Calder) just outside Barnsley, had a cut which went directly to Elsecar (**26**).

Possession of effective transport communication enabled Earl Fitzwilliam to install a Newcomen pump profitably at the colliery and thereby exploit the Barnsley nine-foot seam. In 1812, working at thirteen strokes a minute, drawing forty gallons of water to one stroke, it could drain off 748,800 gallons in twenty-fours hours to release, as a result, 145 acres or four million tons of coal. We see from this that the presence of improved communications, particularly the existence of a convenient canal, was a factor which determined the decision to sink a deeper Elsecar pit and to use a Newcomen engine.

The history of canal development (**27**) is intimately connected with coal mining, not only in this area but in the country as a whole. Exploitation of a coal-mine was the motive for building the famous

19

The Industry

Bridgewater Canal, and the preambles to many late eighteenth-century navigation acts make specific mention of the advantages which would accrue to the coal trade once a positive decision to develop a waterway was made. When the first Duke of Bridgewater said that 'a navigation should always have coal at the heels of it', he was stating what most eighteenth-century colliery owners and canal sponsors knew. Coal was by far the most important canal-carried commodity, so much so that of the 165 canal acts passed between 1758 and 1801, ninety had coal carriage as their prime objective. The reason for this is obvious. Canal transport, before the development of railways, was cheap and unhampered by serious competition. The situation in Birmingham at the beginning of the nineteenth century illustrates this. Before the opening of the Wednesbury Canal, coal had to be brought in by pack horse and waggon. The inconvenience is not difficult to imagine. Lines of horses held up traffic, the heavy coal waggons ripped up road surfaces, and yet produced high transport costs of 13s a ton. The canal reduced this to 6s 8d, and later the system prevailed whereby a ton of coal could be carried any distance over two miles on the Birmingham canal network for as little as threepence. This advantage had been recognised from the beginning of the canal age. From the mid-eighteenth century onwards a fall in the price of coal usually followed the opening of a navigation. It happened in Lancashire in 1761 when the first dead water canal of England, the Bridgewater, was opened. There the waterway, which linked the Worsley mines [**doc. 1**] with Manchester, allowed the average price of coal to be reduced by half. It is so well known that the need successfully to market Worsley coal led to the building of the above canal that the importance of coal exploitation as a promoter of earlier waterway improvement schemes is often overlooked. Yet earlier waterways had been built for coal carriage. Acts for making navigable the Lancashire rivers Mersey and Irwell (1720), the river Douglas (1727) and for building the Sankey Brook Canal (1757) were sponsored by coal interests, and in Northern Ireland the eighteen-mile-long Newry Navigation was opened in March 1742, so that Tyrone coal could be cheaply carried to Dublin. However, it was the Bridgewater Canal which attracted attention and popularised this mode of transport. From 1761 onwards many canals were expressly built to carry coal (**26**). Derbyshire coal travelled on the Cromford and the Chesterfield canals, Midland coal on the Staffordshire and Worcestershire Navigation,

The Growth of the Mining Industry

Wiltshire fuel on the Somerset Coal Canal, and coal from South Yorkshire on the Barnsley Canal. Waterways allowed inland steam coal collieries in the Neath Valley to develop following the opening of a Welsh canal in 1791, and the western sector of this South Wales field came into prominence when other canals opened in the same decade. The Monmouthshire (1792), the Brecknock and Abergavenny (1793), the Swansea (1794) and the Glamorganshire (1794), which linked Merthyr Tydfil with Cardiff, allowed this area to become an important industrial region by providing an amenity which carried in limestone, ironstone and coal and took out founded metals. Where ironworks were small fuel could be transported in waggons or on the backs of ponies, but the large factories employing hundreds of men, which appeared in Birmingham and South Wales, relied on water carriage and consequently crowded the canal banks. Illustrative of this are the 1,500 collieries and ironworks which faced onto Birmingham's extensive waterway system and from which plied annually 68,000 long boats carrying well over two million tons of coal. Colliery expansion and canal mania were clearly related.

The Worsley workings, with forty-two miles of underground waterway, the arterial Bridgewater Canal, and a steady market, enjoyed advantages rarely found in other mines. One of these was the close proximity of transport feature and colliery working-face. Cut coal could easily be placed in runner-shod baskets [**doc. 2**], pushed to the canal barge, loaded into metal containers and sent along to the Deansgate wharves in Manchester. In most mines carriage was a more complicated and difficult process. The problem of linking mine with waterway led to the development of railways which, in both their formative and heroic phases, owe a great debt to the vision of coal-masters and colliery engineers.

The first documented English railway carried coal. It was a simple track built by Huntingdon Beaumont between October 1603 and October 1604 on Sir Francis Willoughby's estates at Woolaton, near Nottingham (**28**), and consisted of rectangular frames of wood laid end to end over roughly levelled ground so that horses could drag waggons with ease. The second line at Broseley, Shropshire (1605), was also built by a colliery owner to link his mine with the river Severn and when, soon afterwards, Beaumont, 'a gentleman of great ingenuity and rare parts', developed others in Northumberland, they were also for pits. From these early beginnings adoption of railway communication was rapid and widespread especially in

The Industry

County Durham (**47**). There generations of engineers tackled basic problems of track, crossings, points, inclined planes, staithe loading and waggon design (**31**). The lines, which worked as feeders between pit and river, were so profitable that by 1750 there were few important collieries without such communication systems. Called waggon ways, and later tram or dram ways, their Tyneside popularity is commemorated in the alternative 'Newcastle roads' [**doc. 2**]. Bertram Baxter (**29**) lists thirty-five public or private coal-carrying tramways in Northumberland, and William Casson's 1804 (**32**) map of the area shows an elaborate network, some of the lines being over ten miles long, winding down to the Tyne and Wear.

In other parts of the country it was canal companies, rather than the colliery owners, which took the initiative and built tramways. These horse lines were cheaper than artificial waterways where the landscape was hilly, and consequently they became an important part of the canal-rail system. This is very true of South Wales, for in Glamorganshire, where mountains limited canal development, the Neath Canal Navigation Company and the Swansea Canal Company were particularly active tramway builders in the first decade of the nineteenth century.

Be the sponsor a colliery owner or the board of a canal company, the commercial motive for developing such transport features was invariably the exploitation of coal resources. Tramways were little used outside the coalfields. A Hertfordshire brick-maker might build one, a short line might be developed to link a chalk quarry with the Thames in Essex, and a few might carry sand and limestone, but in general such railways were exceptional, and the most active patrons of this mode of travel were men of the coal trade. Even the transportation of bulk iron, an important reason for railway development in South Wales, was totally overshadowed as a reason for opening tramways. Of the 538 lines listed by Bertram Baxter in his pioneering gazetteer, over 70 per cent were solely or partially developed to carry coal.

Although iron-railed tramways were well established by the beginning of the nineteenth century and a miner poet could justifiably sing:

God bless the man wi' peace and plenty
That first invented metal plates.
Draw out his years to five times twenty
And slide him through the heavenly gates,

The Growth of the Mining Industry

they had shortcomings. Since one horse was employed to lead each loaded waggon it was an expensive method of carriage. Commentators in 1812 at the opening of the Middleton rack railway made great play of the point that the locomotive allowed the colliery to dispense with the services of fifty horses with the subsequent saving of many tons of animal feed, and there is little doubt that wayleaves which carried horse-drawn waggons could not be very efficient. Junction and terminal points were bound to be congested if, as on the Derwenthough Main railway, 400 waggons traversed the line in a day. Horse-power, counter-weight, or low-powered stationary engines might be used ingeniously, but fundamentally they had limited application. The development of locomotives changed this.

The inventor of the first successful moving steam engine was Richard Trevithick, but the development of a commercially successful line was the achievement of coal viewer John Blenkinsop and Leeds engineer Matthew Murray. In 1801 Blenkinsop was placed in charge of Middleton Colliery and thus became responsible for a railway which already had an important place in transport history. For it was at this colliery that, in 1758, Charles Brandling opened the first railway line to be authorised by Act of Parliament. In return for successfully guaranteeing delivery of 240,000 corves of coal at a maximum selling price of $4\frac{3}{4}d$ a cart, Brandling secured permission to build a line to Leeds. Before the opening coals had sold in the town for $7\frac{1}{2}d$ a corf, and it is interesting to note that the subsequent fall in price is directly proportional to that secured in Manchester in 1761, following the opening of the Bridgewater Canal. When the rack railway opened in June 1812 it was, therefore, but a further compliment to the foresight of this dynasty of Northumberland (**30**) and Yorkshire coal-owners.

Blenkinsop had obtained a patent in the previous year for 'certain mechanical means by which the conveyance of coals, minerals and other articles is facilitated', and when at last the engine did move it was powerful enough to drag twenty coal waggons along the three miles of rack railway to Leeds. The 'Salamanca' was of unusual design. Pinion wheels projected beyond the standard wheels to mesh with a special track. It was an immediate success and if the *Leeds Mercury* tended to emphasise the saving in corn brought about by this replacement for fifty horses—it was the period of the continental blockade—the full significance of the occasion was not totally lost. The crowds of spectators were celebrating the birth of a line which

The Industry

was clearly a viable commercial proposition and, despite the cost, rack railways were immediately adopted at Kenton, near Newcastle (1813), and Wigan (1814).

At the time when Murray and Blenkinsop were working in Yorkshire, other engineers were designing locomotives in colliery yards in the Newcastle area. Within two years of the opening of the Middleton line, William Hedley and thirty-two-year-old George Stephenson had built successful locomotives at the Wylam and Killingworth pits (**31**). Hedley disproved the popular theory that a heavily loaded smooth-wheeled engine would not move forward without a toothed wheel by developing the 'Puffing Billy', which in 1813 pulled eight loaded waggons at five miles an hour, and Stephenson remedied the irregular action of the single cylinder with flywheel by building the two-cylinder 'Blücher'. William Brunton's 'walking engine' and William Chapman's self-hauling chain engine were also being used on north-eastern colliery wayleaves.

In 1819 Stephenson embarked on an even more ambitious project. Using locomotives, stationary engines and self-acting inclines, he developed the complex eight-mile-long Hetton to Sunderland line. This coal-carrying railway was a direct predecessor of the Stockton to Darlington (1825) and the Liverpool to Manchester (1830) lines, and yet was within a tradition of colliery tramway development. What seemed innovatory and spectacular was in reality part of a continuous evolutionary process which went back for at least two centuries. By this time the Killingworth colliery workshop was the centre of locomotive experiments, which enabled Stephenson to develop the direct drive from piston to wheel, slide valves worked by slide eccentricities, spring-supported boilers, malleable iron wheels and the other technical features which make his name a byword for efficient railway engineering, but we should not forget the many other masters and colliery engineers who, starting with Huntingdon Beaumont, realised the commercial potential of railways and solved with ingenuity many complex technical problems.

Coal marketing gave rise to steam railways, but as the century progressed the situation reversed and it was railways which stimulated the coal industry to a growth point far beyond that forecast by the most visionary eighteenth-century *entrepreneur*. Railways changed everything. Small fields, like that centred on Ingleton, withered because coal could be brought in from other areas to compete on

The Growth of the Mining Industry

the local market. Masters' associations found it difficult to restrict wages or prices when potential sources of coal lay beyond their control along railway lines. Tyneside ceased to hold the London monopoly as the South Wales, Midland and Yorkshire collieries sent in overland supplies.

This growth of a coal-burning railway system not only pushed out the boundaries of the fields and forced the search borers to go deeper, but also provided an elaborate transport network which brought coal into easy profitable circulation. The amount of coal carried on the North Eastern system alone confirms this for it had increased from 2,909 thousand tons in 1854 to 13,245 thousand tons in 1880. A comparative study of the amounts of the fuel brought into London by three modes of transport: sea, canal and railway, likewise indicates the importance of the latter form of carriage in the late nineteenth-century economy. In 1834 2,078 (**33**) thousand tons came to the capital by sea and two thousand tons by canal. In 1846, the first year that a figure exists for rail carriage, eight thousand tons came in that way, compared with 3,392 thousand tons by sea and sixty thousand tons by canal. Thereafter canal-carried tonnage dropped dramatically. In fact only a year later, in 1847, the canals carried two-thirds less coal than in 1846. At the same time rail carriage increased. By 1867 the railways had surpassed their greatest rivals, the coastal shippers, and in 1880 6,196 thousand tons reached London overland compared with 3,714 thousand tons by sea. At this time the Midland Railway was the prime carrier to the capital. It transported 2 million tons compared with 1·5 million tons on the London and North Western and just over 1 million tons on the Great Western.

The railway companies were also good customers, for coal not only drove trains but was used to fire the furnaces which produced the 20,000 miles of track which existed in 1900, to bake the bricks for the railway towns which mushroomed at Crewe, Swindon and Normanton, and smelt the metal for the great locomotive works at Doncaster and Darlington. Whereas in 1869 railways were only using an estimated two million tons of coal, by 1900 the figure had increased sixfold.

Railway companies were good customers but the coal-merchant had other patrons. The adoption of power loom and steam-driven machinery in mill and factory meant that the demand for coal further increased. At the same time domestic fuel consumption rose

The Industry

as the population figure advanced. From the time that William Murdock illuminated the façade of Soho Foundry in celebration of the Peace of Amiens (1802) gas lighting ceased to be an experimental novelty and created a totally new use for coal. By 1823 fifty-two English towns were gas-lit and in 1859 there were nearly a thousand gasworks. As with gas so with shipping. The importance of sailing ships was undisputed before the steam tug *Charlotte Dundas* was tested on the Clyde in 1801, but from then onwards coal-driven ships competed with, and finally ousted, sailing ships. This revolution was of great significance to the growth of the mining industry, for steam ships were more than coal-burning machines: they also carried the coal which was being exported all over the world.

The production balance of the various British coalfields was significantly altered because of steam shipping. The South Wales coalfield, which had not been prominent in the eighteenth century, grew in importance because of the appearance of coal-driven ships. After the deep seams of steam coal in the Rhondda had been proved in the 1850s production increased. A series of naval fuel suitability tests, in which the Welsh coal was compared with Newcastle coal to the detriment of the latter, made it a popular fuel. The South Wales fuel lit easily, blew steam up rapidly and, in contrast to that from Newcastle, which left great cokes of clinker and gave black smoke, made little ash. 'I saw stokers throw continuously coal into a furnace,' wrote one observer, 'and when I looked at the funnel I saw no smoke.' The sailor on the Players cigarette packet seemed oblivious to the belching smoke of the flanking ironclads, but English admirals did not want to betray their presence below the horizon and therefore preferred the Welsh coal. Other navies followed Britain's example and consequently the big coaling stations which developed along the main shipping routes of the world received great quantities of coal from ships which emerged from the Bristol Channel and Tyneside. The 'dirty British coaster' was seen by Masefield with some justification to rival the economic splendours of Phoenician trading vessels and Spanish galleons.

The ports which faced onto the Atlantic—Cardiff, Swansea and Newport—expanded because of this trade, but although South Wales had the lion's share of foreign cargo shipments, the North Sea ports also competed. In 1900 42 per cent of the total tonnage came from the Bristol Channel ports, as compared with 30 per cent from the north-eastern, 9 per cent from the Humber and 17 per cent

The Growth of the Mining Industry

from the Scottish ports. Scandinavia and northern Russia, having insufficient coal deposits, were good customers, whose presence benefited Newcastle, Shields, Sunderland, Hartlepool, Hull and Grimsby. Exports were increasingly becoming an important factor in the development of both the shipping and coal industries. From an annual total output of 128 million tons in 1873, twelve million tons was sent abroad. In 1900 the comparable figures were 225 million as against forty-four million. These figures are a far cry from those of 1831 when 237,657 tons of coal, cinders or culm were exported or from 1325 when the first recorded coal ship off-loaded at Calais.

At the turn of the century output figures were reaching their peak. From the estimated 2 million tons produced in 1660 they had risen to 10 million tons in 1800, 64 million tons in 1854, 126 million tons in 1874 to 225 million tons in 1900. Though still a little way from the record year 1913 when over 287 million tons were produced, these figures illustrate an extraordinary development. Complementary with increased production was a new inter-coalfield balance of resources. The output of the traditional Tyneside and Wearside areas was being challenged, and by 1900 more coal was won in the fields Lancashire-with-Yorkshire than in the North East. Yet although such change was gradually becoming inevitable, few guessed that by 1969 all but a handful of pits in Northumberland would be closed.

27

3 Mining Communities

SOCIAL CONDITIONS

When trying to assess the quality of life in a nineteenth-century colliery village emotion easily creeps in to colour judgement. The black and white engravings in the Children's Employment Commission Report (1842) [**doc. 3**] (**36**), which show children chained to heavy sledges or carrying half-hundredweight baskets up ladders, come very readily to mind. The picture of a ragged old woman working a windlass, which lowers a half-naked girl bestraddled across her boy companion into a mine, has been a common illustration in social history books for at least two generations and extracts from Government 'blue books' have also been extensively used. Statements like 'the bald place upon my head is made by thrusting corves', or the eight-year-old trapper's 'I have to trap [open ventilation doors] without a light and I'm scared ... Sometimes I sing when I've light but not in the dark; I dare not sing then', have become commonplace [**doc. 9**]. However, without wishing to support Lord Londonderry's view that nineteenth-century child miners were 'cheerful and contented' (**37**), it does seem important to make qualifications which may modify this picture of unbroken gloom. In the first place it is important to recognise that in many areas, although underground work was dangerous and extremely arduous, the miner enjoyed a relatively high standard of living. Working conditions should be considered separately from living conditions. The point should also be made that where child labour was extensively used extremely harsh conditions and long unbroken hours of labour were new features, which accompanied expansion into large manpower units. Children and women worked in pits during earlier periods as part of family teams, but then the pace was leisurely and included extensive rest periods. There 'Saint Monday' and 'Saint Tuesday' sanctified very regular holidays.

Those contemporary commentators, who did more than write unemotional records of statistics or report on colliery techniques,

Mining Communities

took opposing extremes of attitude when describing life in mining communities. Cobbett is an optimist and Engels a pessimist. When he visited the Sunderland area in 1832 Cobbett allowed that, although nothing is pretty, 'everything seems to be abundant'. The people lived well with free rent, fuel and medical care. Miners' furniture and houses were good, for they took home twenty-four shillings a week [**doc. 10**]. Engels, on the other hand, places a different emphasis. He concedes that miners were relatively well off when compared with factory workers, but dwells at length on illness, life expectancy, abuses of the 'truck' system (see page 31), strikes, child employment and accident (**39**).

These are only two of many writers, but reference to their description illustrates the difficulty of getting to the truth [**doc. 9**]. Couple this problem with recognition that there are seven major and many minor fields, all of which expand and contract at different rates, have varied histories, emerge because of multiple economic pressures to mine various kinds of coal, and the danger of generalisation immediately becomes apparent. Yet taking all this into account, it is probable that Cobbett comes near to the truth about the early nineteenth-century miner's life when he states: 'their work is terrible to be sure . . . but at any rate they live well'.

Mining settlements were often communities apart. Of course some pits were within the boundaries of major towns, but many were sunk in remote areas. This isolation shaped both personal attitudes and community development to make the miner, who at the beginning of the century often resembled an agricultural worker rather than a factory hand, into a conservative. His work contract was traditional [**doc. 5**]. The binding of ploughman to farmer was in essence the same as that which tied pitman to coal-owner. The yearly agreement with obligations on both sides, sanctioned in a past which did not recognise differences between husbandman and collier, was accepted by both sets of workers at the beginning of the century because it was traditional. Miners' roots went deep and this consciousness of a past expressed itself in many ways, but most clearly in superstition, folk lore and song.

If we study our folk song tradition it becomes clear that although nineteenth-century industrial capitalism did not completely stifle growth, it frustrated development so that although the songs of Lancashire cotton operatives and Sheffield grinders are known, they make up only a small part of this heritage. Factory workers, perhaps

The Industry

because they are a product of the recent past, lack a wide oral tradition but colliers, living under different conditions and having a long history, do not. Together with soldiers, sailors, students, and farmers, they possess a communal culture and in the colliery villages of England a continuous music tradition has survived (**41**). As is to be expected, since its history as an organised mining centre is longest, the song tradition of the North East is strongest. Mining ballads are known in published form from the early eighteenth century, but the bulk of survivals come to us from the nineteenth century [**doc. 6**]. Jane Knight of Wingate published a ballad on pitmen's grievances during the 1844 strike; George Walker of Durham printed a song collection in 1839. Song improving duels were a feature of colliery village life up to the end of the century. Like the Anglo-Saxon scop, the pitman poet celebrated the tradition of his community; he was the keeper of the word hoard, who sang of disaster, accomplishment, life, grief and death. 'When you're the Pitman's Poet and looked up to for it, wey, if a disaster or a strike goes by wi'oot a song from you, they say What's wi' Tommy Armstrong? Has someone druv a spigot in him and let oot all the inspiration?'(**40**).

Records of such inspiration give us an insight into mining. The verse,

Jowl, jowl and listen lad
And hear the coalface workin
There's many a marra [mate] missing lad
Becaas he wadn't listen lad,

more properly expresses the danger of underground hewing, telling of the caution and experience required to survive, than does a page of scientific explanation in an N.C.B. teaching pamphlet. Here the common sense of generations is crystallised within the boundaries of a song. This and numerous other traditional ballads and songs of the coalfield are the product of a slowly evolving conservative industry which lacks the sudden invigorating experience of rapid technological change. There is no invention in the history of coal mining before the twentieth century comparable in impact with that caused by the introduction of Cartwright's steam-powered machinery into Lancashire mills. In the mines progress was piecemeal and leisurely, consequently producing a conservative industry. Edward Cowey, an important Yorkshire trade unionist, once told a mining

Mining Communities

conference: 'We are radicals in politics but conservatives in working traditions', and the cultural support for the latter part of the statement is very strong.

A mythology of superhuman underground spirits epitomised by Temple the Big Hewer, a ripper who throws away blunt tools to drill with his nose and cut with his teeth, likewise emphasises tradition. Superstition and the survival at the coal face of words like gate (from the Old Norse 'gata' meaning road) which above ground have lost their original currency, strengthen the impression of conservatism. Other industries have esoteric craft terms, but few trace them to a remote past [**doc. 7**]. Only miners, farmers, sailors, and soldiers express an extensive lore in song, language and legend. Easy communal identification sponsors such cultures, but diversity of occupation does not. For that which is traditional and conservative flourishes best where workers live in isolation with limited occupational opportunities.

The remoteness of colliery villages had social consequences other than the promotion of strong occupation-centred communities. Isolation combined with an irregular wage payment structure to produce in the nineteenth century a widespread use of the 'Truck' or 'Tommy Shop' system of credit sales. Attention has been drawn to the existence of two types of truck (**43**). There was that type which forced employees to accept commodities, usually groceries, at inflated prices in place of coin wage payments, and another kind which, because wages were irregular, made a worker deal with a company store until his earnings were forthcoming. It was the latter which was common in most coalfields, but was particularly widespread in South Wales, Staffordshire and Scotland. The practice of giving 'Tommy ticket' vouchers started at the end of the eighteenth century, increased in the deflationary period following 1814 and continued, despite legislative opposition, into the second half of the nineteenth century.

'Truck' has been seen as an unmitigated evil, which colliers constantly opposed and which even mine owners admitted reduced real wages by 15 per cent. Constant strikes over truck and condemnatory conference motions on the subject strengthen this viewpoint, yet, despite the activity of this opposition, a few points can be made in defence of the system. In the first place not all the owners who opposed the system were as disinterested as they at first seem. 'Truck' was usually a device of small masters to defer payment in

The Industry

times of financial crisis. Owners of large collieries had more sophisticated machinery to tide them over monetary troughs and could thus afford to rail against the iniquities of small-time coalmasters, whose embarrassment would be to their advantage. A Staffordshire coalmaster revealed that the 'opportunity of getting rid of the smaller capitalists who had no other means of carrying on their business', provided a clear inducement to support the 1831 anti-truck legislation. Here it might be well to note Francis Place's dictum, 'Whatever any man who employs a large number of workpeople proposes as a regulation is sure to be something to his own advantage and the disadvantage of his workpeople.' It should also be remembered that in areas remote from a market town the establishment of a grocery store might satisfy a real need and high prices be attributable to transport costs. The presence of credit traders was also useful during seasonal trade depression. Over-production and stockpiling of coal often brought about unemployment. 'Tommy' shops were at such times safety valves which staved off starvation.

However, if the above points can be made in defence of the system, they should not detract from the general impression that 'truck' was disliked by the man and blessed by the masters. In South Yorkshire the enforced taking of tea, coffee, sugar and poor quality meat in place of wages was a grievance which David Swallow, a union leader, exploited at a Sheffield meeting in 1850, when he struck a railing with a piece of board-like meat and declared it 'Tommy shop' bacon (**73**). The case of orphaned Janet James further illustrates the evils of 'truck'. In 1864 she had worked for two years at a South Wales pit pushing coal tubs. Her wages, she said, were six shillings a week but never paid, for they were 'swallowed up in shop bills for bread and tea, on which I lived' (**76**). Her history caused a scandal when cited by W. P. Roberts in the Tredegar trial, and confirmed the already realised injustice of the 'truck' system. The Monmouthshire solicitor, John Owen, giving evidence before a Select Committee on payment of wages, when asked, 'Have you ever known any workmen who were in favour of the truck system?' replied, 'None.'

By the middle of the century many factors which can be used to justify 'truck' disappeared. Communication systems, especially railways, developed rapidly in the 1840s and colliery villages ceased to be remote communities. Copper and silver coin was plentiful so wage payment could be regular. 'Truck' also disappeared because legislation was at last becoming effective and because the men had a successful alternative shop—the Co-operative store.

The notion of Co-operation caught the imagination of the miner in the second half of the nineteenth century in other ways. The Yorkshire Union bought a pit in Staffordshire and attempted co-operative production and the workers of Messrs Brigg's West Riding colliery accepted profit sharing. They received £1,745 in 1871, £5,250 in 1872 and £14,256 in 1873 when they sponsored free co-operation between masters and men. However, it was the presence of the co-operative village store which had lasting effect on the life of colliery communities. Richard Fynes in *The Miners of Northumberland and Durham* (**70**) has described how the West Cramlington pitmen, inspired by Holyoake's *Self Help by the People* (**44**), trudged into Newcastle and on an old cart brought back goods which they sold in a little shop. How after ten weeks £200 had been raised, and in the first quarter almost £450 of goods sold for a profit of £38 15s 10d. Starting from such small beginnings, co-operative trading rapidly spread across the coalfields. Gradually amalgamations took place, wholesale supply became a reality and mean shops were replaced by well-built premises. Co-operative stores destroyed 'truck' by making available cheap quality food and clothing, but they did much more than this. By encouraging thrift and worker-controlled management, they sponsored a democracy which nurtured the sense of personally involved responsibility. The miner and his family owe a great debt to the nineteenth-century co-operative movement.

Another aspect of colliery village life which like co-operative trading induced a sense of responsibility, personal satisfaction and respectability was Methodism. It may well be that its influence as a stabilising force has been over-emphasised, as E. J. Hobsbawm suggests, but the extensive remains of nineteenth-century chapels testify to the existence of vigorous evangelical communities. The influence of Methodists is felt in all aspects of nineteenth-century working-class life, but it is particularly apparent in mining communities. Methodists sat in the miners' lodge, and on Friendly Society committees; they got elected to Local School Boards, became J.P.s, and, after 1884, emerged as a powerful bloc of workingmen M.P.s. They might act as a brake to limit revolutionary progress, be pillars of the Lord's Day Observance Society or Teetotal Association, but be they Primitive, Wesleyan or New Connexion, there is no denying that the Methodists had influence.

The Industry

By the nineteenth century the Church of England had failed to come to terms with the miners, who saw it only as a church of rented pews and tithes which emphasised class. The Oxbridge vicar, who could be both landlord and Justice of the Peace, was an alien. When Ben Pickard fought the Normanton constituency to become its first collier M.P., a supporter, replying to attacks from local Anglican pulpits, asked why dissenters should have to spend double the amount paid by a churchman when they married or pay for consecrated ground when they objected to consecration. This type of criticism struck home, for by 1884 strong allegiance to Methodism and opposition to the established church was a confirmed fact in the West Riding coalfield and other English mining areas. The *Census Report on Religious Worship* (1851–3) points out that, 'the chief dislike which labouring populations entertain for religious services is thought to be the maintenance of those distinctions by which they are separated as a class from the class above them', and in this analysis perhaps lies the root of religious dissent. Independent miners would not tolerate clear differentiation of class if there were alternative modes of worship.

Although Methodism has been singled out as having abiding social importance, it should not be thought that other dissenting groups were without influence. Baptist preaching was popular with colliers and Congregationalism was particularly influential in South Wales. All helped fill the spiritual and educational vacuum which existed in many colliery villages when the Church of England became passive and without enthusiasm. However, it was Methodism, strong in the important Yorkshire, North Staffordshire and North-Eastern coalfields, which produced the great mining leaders and the vigorous chapel communities and therefore it deserves particular attention.

Robert Wearmouth (**49**) has shown that in a list of eighty eminent late nineteenth-century trade union leaders who owed their career, position and influence to personal religious experience, seventy were Methodists. It is doubtful if John Wesley would have approved. He strongly opposed movements tinged with radicalism and throughout his ministry attacked them. It is therefore unlikely that he would have sanctioned organisations aimed at restraint of trade or even condoned moderate Methodist union leaders like Thomas Burt. Early Wesleyans excommunicated trade unionists, and a poster issued in Sunderland in 1810 by a meeting of travelling and local

Mining Communities

preachers and addressed to those members of Methodist Societies who had refused to fulfil their engagements in the collieries, illustrates the vigour of the opposition [**doc. 7**]. In paternalist language it reproaches the strikers, asking if the strike 'is in direct opposition to the command of God' and if they are 'not making work for repentance in their last hours'. Disapproval is clear. Morgan Phillips' belief that 'Socialism in England is Methodist, not Marxist' might be true, but it is a different Methodism from that preached by John Wesley. Wesley could visit Bristol miners and be appalled by their social condition, yet his way of alleviating hardship was not by agitation or strike: it was through prayer. His reluctance to become involved in the worldly aspects of social advance has been criticised; however, it should not be thought that in the long term he was without influence.

What the Nonconformists, and Wesley's followers in particular, gave to the miners was religious enthusiasm and hope. A Commissioner for Mines in 1842 recalled seeing, during a cholera epidemic, 700 men kneeling for an hour in prayer on the pit bank listening to the words of one of their leaders, and the scratched words on a tincan found by the body of a victim of the Hetton Pit disaster in 1860 gives the impression that, far from being a drunken, irreligious rabble, many miners were convinced, enthusiastic Christians. The last message reads, 'Dear Margaret, There are forty of us all together at seven. Some were singing hymns but my thoughts were on my little Michael. Be sure and teach the children to pray for me.' It may well be that religious enthusiasm reached a peak at times of disaster, but the evidence of chapel building and bible ownership indicates ordered, systematic zeal. When Seymour Tremenheere reported on the villages of those coalminers and ironworkers who had marched on Monmouth in 1839, he visited 1,500 houses and inquired about reading habits. Three-quarters of the homes visited had bibles; hymnals and religious tracts were common, but other reading material was scarce. Of 200 householders questioned, only ten produced any other type of book. There is little wonder therefore that by the end of the nineteenth century the language of the bible had become the language of mining conferences. Delegates swopped texts, local leaders urged prayer, and a sure way to discredit an idea was to say that it was the opinion of freethinkers. George Holyoake's atheism was used by opponents of the Cramlington Co-operative (**70**) as an argument against the store, though

The Industry

in this instance the obvious benefits of the co-operative ideal expressed in *Self Help* outweighed the knowledge that the author was in prison for blasphemy.

A group which was very influential in the northern coalfield was the Primitive Methodists. Thomas Burt, the first coalminer to enter Parliament and a powerful Northumbrian trade unionist, was of their number and they were strong in northern mining lodges. Theirs was a proletarian ministry relying on laymen who preached with direct emotional fervour. These preachers sympathised with trade unionism and there is even a case of one congregation, during the long 1844 strike, making supplication and asking the Almighty to act on its behalf and injure blacklegs. Frequently the influence was more constructive. William Brown, an early Yorkshire leader, was more likely to suggest that his men 'give up drinking and start thinking, give up swearing and begin praying' than he was to summon down divine wrath.

Education was a natural correlative of democratic religion and the systematic processes of Methodism benefited many people by emphasising 'self-improvement'. Class-leadership, stewardship and eventually trusteeship engendered a sense of responsibility; the bible class and local preacher's examination promoted systematic education and the lay preaching circuit sponsored oratory. What the Ragged Church Movement sought to achieve, namely congregations dominated by 'persons lowest in the social scale', the Nonconformists managed quite naturally. The chapel in the isolated village provided an opportunity for government by poor people. The coalmasters need not be part of that chapel life, since they often lived, worked and worshipped elsewhere. The manager and his family might attend, but frequently the village chapels were the autonomous preserves of working men. Colliers thus learnt the lessons of administration in chapel and union lodge. Subscription collection, double entry book-keeping, banking and the provision of a building fund became part of both organisations, since both became institutionalised at the same time. The frock-coat and top hat respectability of the union leaders Fenwick, Burt, Pickard and Wilson was chapel respectability. All were Methodists as well as the most influential mining leaders of the 1890s.

It would be a pity if the emphasis placed on Nonconformist religion as a general educating factor caused us to overlook other influences. The propaganda of the Society for the Diffusion of

Useful Knowledge reached some miners, and although the Mechanics Institute movement was town based, in the second half of the nineteenth century its reading rooms appeared in colliery communities. Nor should the contribution of the Church of England be overlooked. National Schools, the S.P.C.K. and the Sunday School movement played an important part in advancing rudimentary literacy. It was within the context of the church schools that the benevolent paternalism of some coal-owners made itself felt. Generally those masters who showed any interest in the 'moral improvement of their workmen' made donations to the local National or British School, but did not teach or superintend classes. The 1842 Commissioners commented on this when condemning 'the employers of labour [who] often hold themselves from moral obligations of every description towards those from whose industry their fortunes spring' and for contrast described the Flockton colliery school. Just outside Wakefield a Unitarian family, the Milnes, not only provided a school-room but also taught miners. Every Sunday they conducted bible study, hymn singing and prayer before instructing the children of their workers in reading and spelling. On Monday afternoons at 3.30 classes for writing, arithmetic, geography, grammar, dictation, 'lessons on objects', drawing and composition met and in the evening a room was supplied with dominoes, chess and periodicals. A Cottagers' Horticultural Society, a choir and a 200-man strong Temperance Society was encouraged. Membership of the latter could be intimidating for violation of the pledge was 'followed by instant dismissal from employment'. In this, as in the compulsory savings system which required every child to save a fixed weekly amount which was only repaid on marriage or movement from the area, these owners, from a modern viewpoint, seem paternalistic. The authors of the Children's Employment Commission saw them differently. To them they were 'the flattest practical contradiction to the asserted inaccessibility of the poor to kindly and civilising influences; and equally so to the doctrine that refinements and labour are incompatible'.

Generally, however, the combined efforts of chapel, school and Mechanics Institute were unsuccessful. Literate trade union leaders emerged but the mass of the miners were untouched by education. The passion for knowledge was not widely distributed with any class and the brutality of a miner's life made the struggle for better conditions so much harder. A determined frontal attack which went

The Industry

beyond sanguine philanthropy was needed if ignorance and squalor were to be overcome. Such a solution required time, energy and money, and the majority of owners did not see it as their problem.

The record of owners as builders of houses is more impressive. Against a background of ill-planned industrial conurbation the houses of many miners seemed well ordered, healthy and clean [doc. 10]. Even from the eighteenth century when the differentiation between miner and agriculturalist is difficult we have descriptions of living conditions which are good. As early as 1768 Arthur Young described an owner's attempt to help South Yorkshire miners by enclosing patches of moorland so that the men 'could raise corn instead of buying it'. Eventually this policy produced a community of farmsteads which was 'a seminary of industry and a source of population' (73). Evidence given to the Poor Law Commissioners in 1842 and to the Children's Employment Commissioners in the same year likewise emphasises that the general standard of housing in pit villages was high. Well-ventilated houses standing in rows or squares with a pigsty and garden patch adjoining are cited. If imperfectly drained common privies did present a health risk, at one place at least miners' wives obtained indoor hot water. At Chester-le-Street, County Durham, water from the engine boiler waste pipe was supplied to the inhabitants of cottages adjoining the workings.

Records of such genuine prosperity are not so uncommon as to seem exceptional but of course there is a darker side to the picture. Well-constructed houses if built by coal-owners were not always an unmixed blessing. Tied housing placed the miner's family under obligation to the colliery master and eviction was a threat which some owners, like Lord Londonderry (54), would evoke when faced with a strike. Then of course not all houses were well constructed. In many areas descriptions of sanitary neglect and squalid housing far outweigh accounts of prosperity, a fact well testified in the reports of late nineteenth-century medical officers of health. The effective sewerage and the public wash houses which miners enjoyed at the Gwthe collieries of the Ebbw Valley were a rarity. Commissioners and the Select Committee reports might urge that owners show more concern about welfare, but in the existing circumstances this seemed a pious hope, for the group loosely termed 'coal-owner' so straddled society that a realistic dialogue with government was almost impossible. One man's resources were not shared by his neighbour. The Duke of Devonshire, because he controlled vast agricultural as well

Mining Communities

as mining property, might be expected to be concerned with the education of pit children, but the self-made master, who was fighting to keep a delicate profit margin, could not. A Welshman working a day-hole which abutted the territory of the great iron and coal magnates, the Crawshays, did not have their problems. There were many sizes of pit and many classes of people involved in mine ownership, and any attempt by Government to speak to or of them, as if they were conscious of general affinity, was senseless. The heads of great houses and men who barely had voting qualification could be termed coal-owners.

The reason why the coal-owners came from such different backgrounds concerns a fundamental principle in property law. In England, unlike the Continent, mineral deposits other than silver and gold are generally the property of the owner of the land surface above the seam. This meant that the great northern nobles received coal deposits, although they were probably uninterested in them at the time, when they acquired land. This acquisition did not mean that they worked the coal themselves. Some did through the agency of their steward but others sub-let the coal seam almost in the same way that they promoted tenant farms. There were of course contract variations, for inevitably ownership qualification is the subject of a great deal of common and statute law, and a multiplicity of owners ensued. The various owners also found themselves restricted in operation. Some early fourteenth-century leases, seeking to provide a definite relationship between the amount of coal worked and the rent paid, contain clauses which seek to limit output. A Whickham lease (1356) says miners must not draw more than twenty tons a day; later the Earl of Newcastle's managers allowed only nine months' work a year and a maximum labour force of seven men. By the sixteenth century such practices were generally disappearing and being replaced by rent levies based on the acreage of coal extracted or the numbers of hewers employed. Later this system of royalty was usually replaced by a fixed rent irrespective of output. In all this we see the reason why by the end of the eighteenth century great regional differences were already in being and why the term coal-owner could be applied to men from many stations of life.

At the one end of the ownership scale there were the coal aristocrats, men like the Lambtons of Durham who, because their vast estates overlay coal beds, combined great economic with concentrated political strength. At times of controversy they were the

The Industry

leading spokesmen for the coal trade in the Lords. In South Yorkshire likewise colliery development was dominated by a few very powerful landlords (**51**). The Duke of Norfolk and Earl Fitzwilliam had influence through a great concentration of pits in the Sheffield area. This had advantages, particularly in the nineteenth century, when possession of large areas of land meant that mining operations could be planned without the distraction of exorbitant demands from owners of intervening freeholds for underground or surface wayleaves. The presence of this aristocracy ensured that the mining interest had the ear of Parliament, and his connection between Government, the northern lords and the mining industry persisted into the twentieth century. It was there even after the nobility's claim to real power was eclipsed by the emergence first of the self-made coal capitalist and later the joint stock company.

At the end of the century a new group of self-made coal-owners had entered the industry and dominated development over large areas. Most of these men and their families had other interests beside coal mining. They might, like the Fentons—the coal kings of the West Riding—have started out as yeomen farmers and developed coal resources, first as a side-line but later as a major concern, while still retaining agricultural interests. Often such family assets extended beyond the coalfield of their origin as did those of the Fentons; for they not only took an interest in coal and canal development from their Yorkshire base, but also promoted businesses in Leicestershire and smelted copper in Swansea. This natural linkage of coal and metal resources into a great industrial complex finds clear expression in South Wales, where the smelting furnaces were run by the same dynasties which controlled the pits. In this area at the beginning of the nineteenth century men who had started life with a small mine and a peck of cunning died millionaires. They never, however, completely dominated the industry, for coexisting with powerful mining families like the Fentons, Brandlings and Crawshays was a group of small colliery owners. These men lived a precarious existence, in which they entered partnerships, sunk mines, went bankrupt, sub-contracted and embarked upon multiple mining enterprises. A step away from the butty master (**53**) they could, if their luck held, invest and become very rich men. In the days when capital outlay was small they flourished and at the end of the eighteenth century, when the tendency to sink new mines rather than expand existing workings was usual, they made up the majority

Mining Communities

of owners. The mines they controlled were small and rarely had a labour force which exceeded sixty. They did not rely on complicated machinery and therefore could survive in the rough and tumble of the early nineteenth-century economy, but as the century progressed and mining operations demanded sophisticated pumps and elaborate transport networks they were forced under. Since they could not afford to dig the deep shafts which penetrated the limestone cap or install the great winding engines, as the demand for coal increased and seams, which had once seemed inexhaustible, were worked out, the financial structure of the industry changed. Joint stock companies, stimulated by limited liability, produced the capital required to dig ever-deeper mines and control of the mines passed into new hands. With the coming of the railways combinations of masters (52), unable to control increasingly sophisticated transport arrangements and coal prices, grew impotent. The rugged individualism of practical colliery men was becoming irrelevant, for probably by the middle and definitely by the end of the nineteenth century the decisive decisions were made by bankers and brokers in remote board rooms.

WORKING CONDITIONS

At the end of the nineteenth century all aspects of the industry were changing. Electricity, together with automatic coal-cutting machinery, was being used and safety regulations were becoming effective. Gone were the days when the injured came home in tubs and corpses in coal carts. By 1900 Mines Rescue Stations and equipped first-aid teams existed in most collieries. The physical appearance of the workings had also altered. Shafts were frequently lined with cast iron, heavy cages lifted the coal and the main underground roads were well lit and ventilated. Discoveries and improvements were still to be made, but by this time we are justified in thinking of coal mining as a modern industry, functioning through relatively homogeneous practices.

This had not been the case a hundred years previously. At the beginning of the nineteenth century, as we have already seen, there was a great variety of ownership systems, mine sizes in no sense approached uniformity and employment practices varied. Yet it is in the method of winning the coal that we most clearly see the nature

The Industry

of regional variation [**doc. 4**] and recognise the gradualness of technical change.

Broadly speaking, two coal-cutting practices existed in the period under consideration. One was called by a range of names but is usually referred to as 'pillar and stall' and the other is called the 'long wall' system. The 'pillar and stall', called by the Welsh 'post and stall', by Scots 'sloop and room' and by Tynesiders 'pillar and bord', is best realised by imagining a draught-board on which the white areas are the stalls and the black the pillars. This type of working is exploited by pushing outwards from the pit bottom and leaving untouched barriers of coal supporting the roof. It is a costly system of mining, for, having sunk an expensive shaft, the owner must reconcile himself to leaving great quantities of easily available coal untouched. Galloway (**1**) reckons that in one fiery pit where eight-yard pillars were left and only four-yard stalls cut away almost two-thirds of the coal was sacrificed so that 39 per cent could be taken out. Likewise the 'ten-foot seam' which outcropped in the Dudley area was not an unmixed blessing. The great height of the coal was difficult to work and large pillars, often ten yards in diameter, had to be left. Here the threat of roof falls was so acute that the northern practice of hollowing out the pillar was not always attempted; instead the workings were sealed and allowed to cave in. In all areas where 'pillar and stall' methods were used the temptation to rob the pillar was great, and resultant roof falls often killed miners. Knowledge of such practices often reaches us through that very profitable historical source, the coroner's inquest.

Another problem presented by the adoption of pillar and stall rather than the long wall system was that of ventilation. As the workings moved away from the pit bottom the tendency for pockets of stale air to collect became real and in such circumstances it was easy for lethal gases to gather. Elaborate systems of ventilation walls and doors (**12**) became a necessity, so that the passage of fresh air could be channelled to prescribed areas.

Although the pillar and stall method has now almost completely disappeared, it was the general system of mining in many parts of the country right up to 1900 and a Yorkshire colliery manager reckons that it was the only system worked in that county before 1886. In that year when the alternative method, long wall working, was introduced into Yorkshire, ignorance was such that a new manager from Nottinghamshire had to be employed to work the new

Mining Communities

practice. The long wall method meant that, as the workings progressed outwards from the pit shaft, all the coal was taken out. The men advanced along a wide front. Having cut access roads at both ends of the planned face, they chopped out a shallow but long slice of coal and temporarily supported the roof with timber props. This done, they advanced by cutting another slice from the walls of this narrow tunnel. As the coal was taken out the 'gob' or 'goaf' [**doc. 7**], the area which began a few yards behind the advancing wall, was packed with stone and rubble. The roof supports were then removed and the area allowed to settle as the roof collapsed.

It is suggested that this method is directly derived from methods used by metal ore miners who, knowing the value of their commodity, place a high premium on the removal of all the mineral. That cannot be proved, but what does seem certain is that early accounts of the use of this system came from an area where coal is not plentiful and tends to occur in shallow bands. Eighteenth-century Shropshire miners at Shifnal used long wall methods, as did the colliers of a ten-inch seam in Somerset and a seven-inch seam in Dr Roebuck's colliery at Kinneil in Scotland. In these cases the height of the seam was influential in determining the method of extraction and this was always the case. British seams varied in thickness from ones where the collier had to crawl out to reverse his pick, up to the famous Staffordshire 'ten yard' seam, and the techniques for winning coal varied accordingly. Yet be the seam deep or shallow, the one great advantage which the long wall system had over its rival was that of ventilation, for where the plan of the workings was simple the air currents could easily thread a direct path and sweep the face where the men worked.

Between the extremes of the long wall and the pillar and stall systems there were intermediate modes of operation, but whatever method was used to extract the coal in the earlier part of the nineteenth century, the 'butty' system was invariably the effective way of managing the mining operation and payment by measure the usual reward [**doc. 5**]. When the 'butty' received payment for his stall it was not reckoned on weight but on quantity of good coal mined. Only the tallies which came up on full corves of coal were counted when the wage was calculated, and this factor alone stresses the power of colliery owners. Fraudulent practices were common for without someone to overlook on the men's behalf, the determination of when a corf was full and what was good coal lay

The Industry

with management. In one place where a separation of small coal from round coal was the required practice a collier was liable to be fined 2s 6d for leaving a peck of small coals in a tub. The presence of a quantity of dust in a corf could also be a reason for confiscating coal and fining the collier. Nor was payment by weight an absolute guarantee of fair dealing. In a minority of pits where, even before the 1860 Mines Act, payment was by weight rather than measure, the workman could not be sure that the scales were correct. The owners of one northern pit required a 'reasonable notice' of three days before they would allow their workers to inspect the weighing machine. As one miner commented, 'if the Inspector of Weights and Measures were to give a reasonable notice to the shopkeepers that they were coming to examine their weights and machines, when they did go, the shopkeeper would take precious good care to be ready for them and have all things snug' (**70**). Thus we see that before the checkweighman clauses, which allowed the men to appoint their own weighing-machine overlooker, were inserted into the 1860 Act, and before the Normansall case had strengthened the men's legal position, the calculation of payment rested with management. It may well be that in most cases the customs were not abused, but because they were open to unfair practice they were a focus for unrest and grievance. For on one point, as Engels (**38**) shows, the justice of the miner's case is difficult to challenge; if the collier was paid by measure, was it fair that coal should be sold by weight?

Of course miners did not accept this unfairness. Their leaders fought to change the system by calling for legislation and by promoting strikes. Industrial action aimed at altering the payment systems is a common feature of the nineteenth-century scene. The terrible strikes and lockouts which rocked the North-East in 1844 were not isolated incidents. In all coalfields the terms of the miners' contract caused unrest, the system of payment by measure being seen as clear evidence of oppression.

Yet although this unrest existed by general report, the collier was seen to enjoy a relatively high standard of living and to be well paid. Of course we cannot discount the extreme and unique harshness of the miner's life, and any assessment of pay scales must be placed against this background, but, having said this, it is important to recognise that generally the miner's wage compared favourably with other workers of the time. In 1889 Derbyshire lead miners

Mining Communities

worked longer hours than Yorkshire coal workers for less money, and in the same period a 100-hour-week London bus driver earned less than the collier who worked for fifty hours. These illustrations from the end of the century only confirm what had been noted earlier. Even the eighteenth-century Scottish miners, often compared to slaves because of the legislation which bound them to the pits, received, if Adam Smith is correct, three times as much as the common labourer. This tendency to earn more than other semi-skilled workers continued, despite acute trade fluctuations, throughout the nineteenth century. At its beginning Northumberland miners earned 2s 6d to 3s 0d a day, and worked on average a 4½-day week for a weekly wage of something like 13s 6d. By 1900 35s 0d was a common weekly wage and the collier worked an eight-hour day. This statistic should not be taken as a norm but only as a reference point, for throughout the period there existed great regional variation. For example Durham hewers got 7s 6d a shift in 1874 but only 6s 0d a shift in 1900, a reversal of rates which strongly contrasts with those which operated in Yorkshire at the same period. But the main point should not be missed. At 6s 0d a shift the miner was getting quite a good wage, for in 1900 an agricultural labourer was averaging only 15s 0d for a seven-day week.

Even when given in full detail, a wage index based on daily or weekly accounts does not give a true picture of the collier's wage. Bounties on the one hand and fines on the other make a lot of difference to living standards and, more than most workers, miners received perquisites and suffered deductions. Stoppages because of the presence of quantities of dust in the corf and fines imposed because the tubs were underweight have already been mentioned [**doc. 5**]. In addition there were wage deductions to pay for candles and powder. Sometimes these were considerable. Tremenheere in 1840 reported that one collier, who earned £6 10s a month, spent 11s 2d on candles and powder. In another instance a steward considered it one of his perquisites to make the men buy candles from him at 1½d or 2d a pound above the market price. Payment for broken tools also reduced a wage. If the workers of Stainboro' colliery in Yorkshire are to be believed, the payment which the owners demanded was far in excess of the current market value of the goods. For a shovel shaft they were charged one shilling and a peggy shaft cost sixpence 'although hitherto it had cost twopence'. Obviously, by using such methods, an unscrupulous manager could work an

The Industry

extensive fine and replacement system and thereby partly control wages by backdoor methods.

Yet on the other hand it could be argued that because of hidden assets, but not necessarily money payments, the collier was very well off. His free coals and colliery house, despite the obvious disadvantages which always accompany such rewards, must be taken into account when assessing a standard of living. Also the grants of ale, bread and cheese, which accompanied any departure from a routine schedule, should be remembered. Illustrative of such customs are the gifts to Lord Fitzwilliam's 1,100 workers, many of them miners from the Rotherham area, who received a money present and a handsome piece of beef on St Thomas' day and a piece of bacon on 3rd February each year. Work incentives in the form of money payments were also given to many miners, both to encourage regular attendance and to increase productivity. Miners in Rothwell in Yorkshire in 1816 received four shillings, called 'takking brass', when they worked a full week of six days, and similar output bonuses are met with elsewhere.

The case concerning the reality of a high standard of living is open to dispute. What cannot be questioned are the dangers inherent in mining, and if the miner was overpaid compared with other workers, he deserved to be for as Engels remarks, 'in the whole British Empire there is no occupation in which a man may meet his end in so many diverse ways as in this one'.

When thinking about the dangers of mining, it is easy to emphasise the dramatic and forget the undramatic dangers which face the collier. To a generation which remembers Aberfan, the horror with which the people of Barnsley heard of the Oaks Colliery explosion and 334 dead in December 1866 is easily imagined, but it is not so easy to picture the suffering of a man dying over a number of years from pneumoconiosis. There is a saying among pitmen that for every ton of coal brought to the surface a pint of blood is spilt. This 'blood payment' is easily imagined as the outpouring of a body gash but we do not easily recognise that it can equally be blood coughed from lungs corrupted by dust. From 1851 onwards Mines Inspectors' [**doc. 11**] reports have tabulated the deaths and injuries from roof falls and explosions. The attempt to compile statistics about nystagmus and silicosis has a shorter history (**58**). Yet in the long run it is probably these and other health hazards which have had a lasting effect on mining communities. Miners and their wives over generations learn with almost stoic logic to accept violent injury, but it is not so easy to live with chronic illness. Rupture, nausea, rheumatism, 'beat knee' and 'inflammation of the joints' join the respiratory diseases, which deriving directly from his occupation, crippled—and indeed still cripple—miners. Old age came prematurely to nineteenth-century miners; 'mashed up at 40 and 45' their life expectancy in 1844 was about forty-nine years. This was ten years below the national average for all men who had survived beyond their nineteenth year. Most died of respiratory diseases and were thus classified as men dying from asthma or that all-embracing Victorian classification, consumption. Forced by the physical nature of their work to take in excessive amounts of air, they were compelled to work in conditions, like those in the red-ash pits of South Wales, where there was scarcely enough oxygen to support the flame of a candle. Managers argued that workers became acclimatised to the conditions but the parish registers tell a different story. The extremely unwholesome conditions of poisonous atmospheres in the long run probably killed more men than all the sensational gas explosions of the North East and the regular roof falls of Staffordshire, yet, since they cannot be classified or analysed, they can only be brought to mind in statements like that of the miners' doctor who declared that he knew of no old colliers.

When we come to look at the other causes of death underground, literature on the subject is more easily available. Many statistics relating to violent death in collieries are obtainable and the ways in which engineers and overlookers learned to overcome ventilation and technical hazards are well documented. The work of Clanny, Davy and Stephenson on safety lamps is fully recorded, as is that of the Speddings, Buddle and Brunton on ventilation. From 1851 the yearly statistics of H.M. Inspectors of Mines have been easily available and, thanks to the recent research of P. E. H. Hair (**61**), information relating to the shadowy period 1800–50 can now be studied. From an early date the problems of violent death and injury have attracted clear-headed and careful commentators.

From very early times colliers have died because of the presence of mine gas. The report in the Gateshead register of 1621, 'Richard Barkas burned in a pit', finds parallels in coroners' inquests and church records throughout the country. Perhaps the greatest difference, however, that exists between these early reports and those which are given in the nineteenth century concerns the number

The Industry

killed. Prior to 1812 the number of people killed in a pit accident rarely exceeded ten, but later catastrophes involving over one hundred dead occur. To a degree this is evidence of colliery expansion, for in earlier periods the disasters at Felling (1812), Risca, South Wales (1860), and Ardsley Oaks, Barnsley (1866), with their respective death rolls of 92, 130 and 334, could not have taken place, as such great numbers would not have been down the same pit.

Colliery expansion therefore was an indirect reason for high mortality figures, and one of the reasons why early statistics are suspect is because an accurate account of the numbers employed in mines at the beginning of the century does not exist. Without such figures comparison is difficult before the publication of the mines inspectorate's reports in 1851. But, although a scientific attitude to this subject is not possible until after that date, two general observations about the conditions which existed between 1800 and 1850 can probably be made. One concerns the causes of violent death, suggesting that there were definite qualitative as well as quantitative regional differences, and the other relates to the belief that, rather than reducing the number of mining deaths, the introduction of the safety lamp increased the danger.

By 1850 in the mines the mortality from violence was four or five times that of the general population and stood at something between 4·5 and 5·0 per thousand. Yet there were striking regional variations both in the numbers killed and in the manner of death. In 1851 the Black Country field was about twice as dangerous as that of the North-East. In the Midlands deaths were not caused by explosions but by roof falls. The difficulty of managing the high roofs killed more men per thousand workers than did those gaseous mines of Yorkshire and Durham, which generally come to mind when one thinks of the dangers inherent in mining. From this we see that improved safety conditions were not only a question of ventilation and improved lamps, but also concerned better propping and a scientific appreciation of strata.

The background to the generalisation concerning the introduction of the safety lamp and mortality can be quickly given. In 1812 an hitherto unparalleled explosion occurred at the Felling Colliery (**32**), County Durham, and ninety-one men, over three-quarters of the labour force, died. Felling had been thought a safe pit and this, together with the magnitude of the disaster, caused the newspapers to promote interest in mining casualties, rather than suppress

Mining Communities

information as previously. The public responded, and in October 1813 the Sunderland Society for Preventing Accidents in Coal Mines was formed. It was known that the provision of a lamp which would not explode the gas was a prime demand. An earlier generation had put its faith in Carlisle Spedding's steel mill, a hand rotary steel which struck a flint to emit a shower of sparks which were not thought hot enough to ignite the mine gas. A series of explosions at the turn of the century had shown that this device did not provide a total solution, for the sparks could be dangerous. Therefore in 1815 the Sunderland Society wrote to Humphry Davy, already an eminent chemist, inviting him to investigate the problem of making a safe miner's lamp. After visiting a gaseous pit, and talking with Dr Clanny whose experimental lamp was already made, Davy quickly constructed a lamp of glass tubes and metal, which was a success. Whether this was the first effective lamp is a matter of dispute. At Killingworth Colliery George Stephenson had already made an experimental lamp, and some authorities consider that this gives his invention a prior claim, suggesting that he came to conclusions by trial and error, whereas Davy's slightly later conclusions were deducted from scientific theory. Other writers challenge this with the evidence of J. H. H. Holmes (**1**), a knowledgeable contemporary of Stephenson, who saw the famous engineer try, but fail, to explain how the air and gases worked on the flame. They express the thinly disguised belief that Stephenson's was an act of plagiarism. Yet be the invention the work of Davy, Stephenson or Clanny there is little doubt that safety lamps were well received by the north-eastern mining community especially after both Davy and Stephenson had tested their lamps and gambled with their lives by venturing with them into notorious 'blowers'. Davy lamps were in use at Hebburn colliery in January 1816, and such was their success, that by the end of the year a pit as far afield as Rothwell Haigh, Yorkshire, was buying safety lamps. From then on the use of such lamps increased, and, although most mines continued right into the second half of the century to use 'mixed' lighting—flame in some sections and safety lamps in others—the underground workings gradually became better lit and safer [**doc. 11**].

The spread of the safety lamp can be traced, but the consequences of its adoption are more complex. It is well known that colliers, ever conservative, sometimes struck when a safety lamp was introduced,

The Industry

arguing that its light was dim and that this, because it limited production, restricted earnings. It is also known that the lamp was misused; that the gauze was lifted when the miner wanted more light or a pipe of tobacco. In addition, however, and this is more important, it is often suggested that in the initial stages of its introduction the safety lamp was a positive danger, for it encouraged management to cause men to venture into gaseous regions which could not have been worked prior to its adoption. Evidence for this is usually based on the 1825 pamphlet *A voice from the Coal Mines, or a plain statement of the various grievances of the Pitmen of the Tyne and Wear* (**61**) or on the *Report of the Select Committee on Accidents in Mines 1835*. The latter, after commenting on the number of lives lost in the eighteen-year period prior to 1816, recognised that in a corresponding length of time after that date more miners were killed. It goes on to say: 'To account for this increase, it may be sufficient to observe, that the quantity of coal raised in the said counties has greatly increased: seams of coal, so fiery as to have lain unwrought have been approached and worked by the aid of the safety lamp' (**59**). It now seems that this could be a suspect conclusion, and that, rather than being a device for getting more coal for greedy owners, the introduction of the lamps had immediate humanitarian effects. As P. E. H. Hair has pointed out, the two comparative sets of figures—477 killed in the eighteen years to 1816 and 538 killed between 1816 and 1834—are suspect. Since no deaths are given by the local northcountryman who compiled the list for the years 1801, 1804, 1807, 1810, 1811 and 1816, we must either conclude that no one died in these years, which seems improbable, or that this list is incomplete and what appears as an eighteen-year span in reality is only one of twelve years. In this case the figures are no longer comparable and therefore conclusions based on them are invalid. In fact another list of casualties, again admitted to be crude and deficient, can be used to point a totally different conclusion and show that the number of deaths per thousand employed decreased after 1819 (**61**).

Coupled with this problem of the consequences of the adoption of safety lamps is the bigger one of mine ventilation. Throughout the eighteenth century the efficiency of ventilating practices had troubled the mining engineer and manager. Obviously, pits sunk to depths of 500 ft with complex systems of galleries presented different problems to small day-holes, and therefore by 1800 a body of empirical knowledge about ventilation was in existence. By the

Mining Communities

middle of the eighteenth century James Spedding of Whitehaven (**1**) had devised 'air coursing' a system which relied on vertical stoppings, gates and partitions to make the air thread through and sweep all the galleries [**doc. 13**]. Likewise the practice of employing a fireman to explode the gas was well known, being mentioned as early as 1672 in Sinclair's *History of Coal Mining*. Thus we see that at a relatively early date, the two fundamentals of the ventilating practices, known as the diluting system and the firing system, were in use (**12**).

The important breakthrough in the history of mine ventilation came, however, when John Buddle devised the method of sub-dividing the workings into a number of independent ventilating systems. This system of compound ventilation with its multiple regulating doors necessitated the employment of door keepers, and consequently that much-written-of worker, the child 'trapper', became an essential labourer in the mines. Compound ventilation was very successful, solving many of the new problems which arose when, in the search for rich seams and quick profits, the management dug deeper and found the fire-damp an increasing hazard [**doc. 12**].

A successful ventilating fan was not in general use before the latter half of the nineteenth century. Up to that time, the primary air circulation was maintained by simply noting that hot air rises and oxygen, by feeding a fire, creates a vacuum. The time-honoured practice of vertically dividing the shaft and then placing a fire bucket or a ventilating furnace at the foot of or in a funnel at the head of the shaft and sucking the air round, was used. The attendant dangers—explosions and underground fires—can be imagined. However, by 1850 ventilating efficiency, even with this type of device, was such that very high air currents could be induced. In 1835 Wallsend Colliery could only maintain a current of 5,000 cubic feet per second but fifteen years later 121,360 cubic feet per second was achieved. By then, however, mechanical ventilators were being used. Inventions like John Buddle's Air Pump (1807), John Martin's Air Lock and Fan (1835), William Fourness' Rotary Air Drum (1837), Benjamin Biram's Fan (1842), and William Brunton's Ventilator (1849) were being surpassed by reciprocating engines acting by displacement, like those of Struve (1846) and Nixon (1859) and centrifugal engines like those of Waddle (1865) and Walker (**12**).

From mid-century improved ventilation, and the introduction of fans in particular, reduced the number of fire-damp explosions but

The Industry

this improvement was partly offset by a new hazard, the coal-dust explosion. Right up to 1900 the question as to whether the dust could explode spontaneously was an open one. Vital, the French engineer who investigated the Campagnac explosion, said it could, as did Dr William Galloway of Cardiff after studying the Penygraig disaster of 1800, but their conclusions did not find general support, even the inspectorate being divided on the issue. In 1887 the Coal Mines Act required that dust be watered for a twenty-yard radius in the vicinity of a gunpowder shot, but the 1896 Coal Mines Act made no reference to watering. Some Mines Inspectors therefore enforced the requirements of the earlier act, while others did not enforce watering, since it was not specified in the 1896 legislation. In such confusion many managers naturally remained insensitive to the dust hazard and consequently miners died because of this understandable negligence. Dynamite blasting was becoming common practice, and, because of the failure to understand the nature of the new threat, in one sense mining operations in some areas became more hazardous. The general death toll was decreasing and production was rising, but in the period 1885–95, if we measure by deaths the severity of accidents caused by explosions, then in broad averages the number of deaths is doubled. Stone dusting was almost unknown, and because of this coal-dust explosions were clearly taking over from firedamp explosions as a principal cause of death in mines.

Yet, although this is so, as the *Colliery Journal* pointed out in its first number in 1861, more continued to die from roof falls than from fires and explosions. In the period 1851 to 1860 the average number of men who were either crushed to death or who were killed when the roof fell in was 377. When compared with 250 who died in explosions during the same period, the magnitude of this unspectacular danger is recognised. Accidents in shafts were another hazard which faced the mid-nineteenth-century collier. Over 200 men were killed annually in the shallow shafts at that period, for going down in an open basket or clinging to a rope as part of a chandelier of descending miners was bound to result in accidents. The introduction of safety cages together with better winding and signalling devices altered this and as time went on, because of these and similar improvements, the rate of accidents decreased. In 1851, 985 men died at a death rate per thousand underground workers of 4·3; in 1871 that rate was reduced to 2·3 and in 1886 was slashed to 1·8. This is a remarkable reduction, particularly when it is recog-

Mining Communities

nised that this period saw a dramatic increase in the number of men employed, for, although the working force had risen to almost three times its 1851 strength, only 107 more men died in 1886 than in 1851.

It is also important to see that such improvement was not achieved accidentally, nor was it solely the result of technical improvement. The improved ventilation systems, the safety devices, the improved lamps invented by Upton-Roberts and Mueseler, together with the stone-dusting techniques of Atkinson, all had a part to play, but without the body of mining legislation many of the improvements would have passed unheeded. Safer and better working conditions are not always of prime importance when weighed against earnings or profits, and the miner was fortunate in having a body of parliamentary legislation to protect him. In the long run the Mining Inspectorate was as important as the technical innovator, for, although some trust could with confidence be placed in management, implementation of laws relating to safety would not always have occurred had not the inspectors had the obligation to report on conditions and the power to fine the recalcitrant.

4 Coal Mining Law

When Lord Justice Mansfield in 1772 ruled that slavery was offensive to English custom his judgement pointed to a serious national anomaly which attracted immediate comment. Men asked why, if it was correct to free a Negro, did Scottish miners (**57**) continue to be restrained within a legal framework which accounted them little more than slaves. This argument is advanced in the preamble to the Scottish Miners Emancipation Act (1774) when it is urged that the grant of freedom 'would remove the reproach of allowing such a state of servitude to exist in a free country' [**doc. 9**].

The accuracy of the equation which related slavery to colliery serfdom can be questioned, but there is no doubt that the social and legal status of the Scottish miner was low. He was a man apart from other men, for when the Scottish parliament in 1641 strengthened legislation passed in 1606 to control the movement of colliers, and made it necessary for a collier to hold a leaving certificate before he could move from his place of employment, it effectively forged an instrument which bound a man to a master for life. The common custom of accepting 'arles', a money or commodity payment, as a token of engagement re-enforced the system of servitude. Arling, in fact, not only bound the man for life; it also tied his children to the coal. A gift at baptism was part of a contract under which a man agreed to bring up his child as a collier. Thereafter serfdom became the boy's heritage, a heritage which sometimes caused him to be listed on a colliery inventory.

Of course the picture can be overdrawn. Scottish miners, unlike many plantation slaves, did have freedom of marriage and religion. They were also well paid, for even the most conservative wage assessments place their income above that of contemporary Scottish free labourers. Yet since many miners, although not all by any means, could not move without permission and because run-away colliers, who were not free for over a year and a day, could be hunted down, their state of servitude seems extreme in Scotland. Coal mining was

Coal Mining Law

not an occupation for free men but an alternative mode of punishment which judges could inflict when they did not want to sentence a man to transportation or death.

The two acts of Parliament which relieved the Scottish miners, the acts of 1774 and 1799, were not totally motivated by philanthropy and justice; they were sponsored because there was a labour shortage and because, as their countryman Adam Smith pointed out, 'the work done by slaves ... is in the end the dearest of all'. The rapid growth after 1760 of Scottish ironworks meant that more men were needed to work the coal-mines. Colliers were not coming forward and therefore the masters sought to remove the social stigma of serfdom by initiating emancipation legislation. The first act (15 Geo. III, cap. 28) was only partly successful, but, because the initiation of the legal proceedings which would free the individual miner and his family rested with each man, many failed to apply to the Sheriff for their freedom. This situation persisted until 1799 when a new act (39 Geo. III, cap. 56) finally ended serfdom by automatically freeing all those miners who had failed to obtain a decree of the Sheriff's Court.

The situation in Scotland was of course unique, for English miners were not bound for life. Their usual contract was a yearly agreement called 'The Bond' [**doc. 5**]. Theoretically this contract was mutually beneficial, for, in areas of acute labour shortage, it guaranteed that a master had workmen who were bound to him from one hiring fair to the next, and it also sought to protect the men by ensuring employment even when coal had stockpiled and prices were falling. Under such a system the laws which were enacted to free the Scotsmen were inappropriate and over forty years of the nineteenth century passed before a major piece of legislation affecting English miners occurred.

At the beginning of the nineteenth century the *laissez-faire* doctrine was dominant. Factory legislation which in any way smacked of State interference was chopped at so that only the most innocuous reformist clauses remained. Under such conditions the relief of the coal-mining communities was slow. The Truck Act of 1817 forbade payment of miners' wages in kind but lacked the teeth to enforce its principles, and, even if the law had been a total success, it was not really the type of legislation which attacked the roots of the coal-mining community's distress. Even an owner 'would go a hundred miles to see a butty [who kept a truck shop] hung', and the Truck

The Industry

Act was piecemeal legislation which looked at one small grievance but left the major areas of affliction (the increasing employment underground of women and children, long hours and insufficient safety regulations) unreported and neglected. This was partly because of the absence of a parliamentary lobby interested in the social welfare of mining communities, for at the beginning of the century the miners lacked an effective advocate who, having an intimate knowledge of the industry, could call for improved working conditions. There was no Peel senior, Oastler, Sadler or Fielden, and, although it is true that John Buddle and others were concerned about safety regulations and ventilation, they shied away from general legislation. The Sunderland Society thought that its work was done when Davy presented his lamp, and thereafter went into voluntary liquidation. The terms of its foundation had been narrow and it was not concerned with wider social problems. In its narrowness it represented typical attitudes which persisted among many coal-owners. For although these men were not dyed-in-the-wool villains who lacked common decency (indeed many were genuine philanthropists) they could not believe that there were general rules of conduct which would govern the workings of an efficient coal-mining industry. The belief that parliamentary legislation was undesirable since it would remove the obligation of the owner to improve his men's situation, though sincerely held by many nineteenth-century capitalists, echoes hollowly in the modern ear, but the argument that working conditions in the various fields were so diverse that a valid list of General Rules would prove impossible to formulate, is understandable [**doc. 4**]. These owners were experiencing a rapidly changing set of circumstances in which output advanced at an unprecedented rate, manpower increased and technology introduced new problems, so there is little wonder that judgement faltered. In such circumstances it was difficult to be totally objective or to see that some improvements would only be forthcoming if Parliament intervened.

However, before Parliament would act it was necessary to show that a problem existed. This was done through the setting up of a number of committees and with the presentation of a series of reports. Some of these bodies were *ad hoc* groups, which, in order to solve specific problems, sponsored lectures and propagated their views in newspapers. Others were formally constituted parliamentary committees which presented full and detailed documents like the *Report of the Select Committee on Accidents in Mines* (1835) and the *Report of the Commissioners on the Labour of Women and Children in Mines* (1842). Into this latter category comes the first influential report to affect miners.

In 1833 the Whig government, having accepted the need for some control over conditions in factories, passed Althorp's Factory Act. This act was the outcome of extensive investigation of working conditions in various parts of the country. One team of investigators, led by A. Carleton Tufnell, reporting on Lancashire became aware that conditions in mines were often worse than those in factories, and when this was confirmed from South Wales by Hugh Tremenheere, one of the first government inspectors of schools, the way was cleared for a thorough inquiry into the mining industry.

Hugh Seymour Tremenheere (1804–90), who in 1840 presented the clear and vivid picture of life in a mining community in Wales which was influential in causing many people to view such areas with concern, was a civil servant who exceeded his brief. If, as instructed, he had just been content to report on the state of elementary education in South Wales then there would have been little enough to say. However, he was not content to do this but analysed instead the factors which stifled education and which contributed to the unrest which culminated in Frost's 1839 rebellion. He showed that 'truck' was a positive evil, gave family budgets which showed that, although many men earned £6 10s a month, much was spent on beer and that in the end this sort of vicious society, because of its hopelessness, was liable when provoked to attack the Mayor of Newport. He revealed the minutiae of the society—the Temperance Clubs which expected their members to contribute twopence a week for beer—yet his canvas was also broad and by being comprehensive prepared public opinion for the wider survey which was presented two years later by the Royal Commission.

At the same time as this government official was collecting his information, in the North East another committee, realising the necessity of having a clear appreciation of the state of the mines, was meeting regularly. The South Shields Committee, motivated by the growing number of people killed in local accidents and particularly by the fifty-two deaths which occurred following an explosion at St Hilda's pit in 1839, was so thorough in its investigation that in the same year it was able to publish a report which recommended three

The Industry

measures which are of fundamental importance. These were the compulsory registration of plans of mines; the institution of a system of government inspection; and the prohibition of the employment of women and children below ground. The necessity of registering plans is recognised when one considers the dangers from stale air, gas and great volumes of water which would beset workers who accidentally broke into an abandoned yet extensive system of underground workings (**64**). The desirability of an inspectorate is also apparent, but it was the statements on behalf of women and children which eventually pricked the public conscience. In 1840, as a result of pressures from such groups a Royal Commission of four members was appointed. Two of the commissioners, the economist Thomas Tooke and Dr Southwood Smith, were veterans of the Factory Commission of 1833, and these together with two factory inspectors and twenty sub-commissioners were authorised to investigate and report, but not to make recommendations about the employment of children.

When the *First Report of the Children's Employment Commission* [**doc. 3**], the report dealing specifically with conditions in mines, was published in 1842 its impact was such that a public which lived outside the boundaries of the coalfields, forcibly awoke to the terrible conditions which existed below ground in many mines and, thus educated, demanded action. The volume of this criticism was partly promoted by an innovation. Southwood Smith, like many other good publicists, recognising that busy people lack the time to read detailed texts, commissioned a set of pictures drawn on the spot by Binney to accompany the report. Those who glanced at the illustrations and those who read the text became aware of the same indignities. They learned that the chain which harnessed a woman to a coal sledge hurt 'worse when we are in the family way', and that a six-year-old-girl carried 50 kg. of coal fourteen times a day on a journey equal in distance to the height of St Paul's Cathedral. One clergyman unwillingly testified that he did not think that the peculiar bend in the back, necessary for efficient employment, could be obtained if the children were set to work at an age later than twelve and the readers became aware of this interesting, if unusual, opinion. Seemingly for the first time, M.P.s heard of the trappers, who crouched in a hole in the gallery wall so that the pit could be ventilated. In clear and concise language they were told of the open sexual promiscuity which existed because naked men worked beside

Coal Mining Law

women and young girls, and the public, ever sensitive to what it considered indelicate, responded with an outburst of indignation which probably could not have been engendered by the knowledge that in notorious Derbyshire 'butty' gangs of children were required to remain in the pit for thirty-six hours when working double shifts. Human misery together with spiritual neglect was shown to exist and the Commons was thus prepared to accept Ashley's bill (37) when he presented it to the House a month after the publication of the report. Such acceptance was a triumph of thorough reporting, for after the presentation of the text it was clear that it was unreasonable to expect reform from within the mining industry. Change cost money, and, since there was no hope of compensation, neither industrialist nor worker was prepared to act. The men dependent upon the labour of their womenfolk and children opposed reforms which would radically alter time-sanctioned work practices. While management, even when ready to curb abuse, realised that, unless reform was universal, the competition of unreformed pits could be destructive to trade, for it would allow prices to be undercut. In such circumstances State intervention was the only solution.

The Mines Act (5 & 6 Vict., cap. 99) which became law on 10 August 1842, although it can be rightly seen as legislation which embodied the principle of State intervention, must also be recognised as a moderate measure. Ashley, even before the bill had left the Commons, had, on the entreaty of John Buddle, Lord Londonderry's agent, agreed to ten rather than thirteen as the age at which boys should enter the pit. In the Lords, when under attack from Londonderry himself, further compromises had to be accepted. The Commons sanctioned the idea that inspectors would only allow these officials to report on the condition of mine workers. The agreement reached by Ashley and Buddle, which said that thirteen-year-old boys should only work on alternate days, also went (54). Unable to throw out the bill, the Lords' policy was amendment and then more amendment.

Yet however timid the act might appear it was the initial step which was needed if the pits were to be reformed by parliamentary action. Women were prohibited from working underground, despite very genuine resultant hardship in areas where there was no alternative employment. In Scotland something like 2,400 women were involved because of Section I of the act, and, although this large figure diminished the farther south one went, there is little wonder

The Industry

that many women continued to work dressed and disguised as men. In Yorkshire in 1843 one female was employed for every forty-five male colliers, but in Midlothian the proportion was down to one female for every three males. Under such circumstances there is little wonder that Scotswomen of Tranent, made miserable by legislation which sought to succour them, should abuse Ashley and all his works.

Section II agreed that boys should be allowed to enter the pit at ten, Section VIII thought that safety would not be impaired if boys of fifteen should act as winding enginemen, and Section III, while authorising the appointment of inspectors, limited their powers to such an extent that their potential good was nullified. All this seems piecemeal legislation until we recognise that major points of principle had been conceded. The necessity of inspection was established, the desirability of Government-authorised safety regulations sanctioned, and the need to give age limits stated. For the first time women were excluded by law from a particular kind of labour. These were very important legal gains and beside them the section forbidding the payment of wages in public houses seems a paltry throw-back to a past of ineffective 'truck' and trade regulation. The 1842 Mines Act was of fundamental importance, because for the first time Parliament, stimulated by a series of inquiries instituted under the growing force of public opinion, directed its attention to social conditions in mining districts, did not like what it found and acted.

The story of the next eight years is of abortive parliamentary action to strengthen the 1842 Act. It is also a history of the application of pressures which in the end succeeded in inducing the Government to present its own act. An Act for Inspection of Coal Mines in Great Britain (13 & 14 Vict., cap. 100) which slipped through both Houses without much trouble in August 1850 embodied a principle which had received half-hearted approval in 1842 with the appointment of inspectors; this was the premise that the State might and ought to interfere, in the interests of safety, with the working of the mines. An experimental measure, initially sanctioned for only five years, the act did not come up to the ambitious proposals of the reform enthusiasts. Nevertheless, it did cause four inspectors to be appointed [**doc. 10**], made sure that plans were kept and produced for inspection when required, and said that fatal accidents must be immediately reported to the Home Office. The inspectorate was also authorised to go underground and, if still without molars, at least it received milk teeth in the section which allowed penalties of £5 to £10 for obstructing inspectors. Not unlike the 1842 Act, this was an act which was in reality the first real step towards an active regulation of working conditions. The 1847 prayer of the recently formed Miners Association of Great Britain that 'Inspectors should be appointed to visit all the mines and that some of these inspectors should be men acquainted with colliery work' was at least partly answered, and the step towards the introduction of practical working-men inspectors was thus much nearer.

The first Mines Inspector, however, was neither a practising miner nor a knowledgeable engineer; he was a young Wykhamist barrister and a civil servant, who had made a small reputation in education. In 1834 it was not deemed necessary to appoint inspectors who were directly associated with the object of their investigations; indeed, it was often considered desirable to appoint men who initially were ignorant of the subject which they were asked to study. The first two inspectors of schools did not teach and the first Mines Inspector had little to do with collieries before he went into South Wales in 1840. Tremenheere's promotion to the office of first Inspector of Mines got the Government out of a delicate situation— Tremenheere as one of 'Russell's Bashaws' of schools had been too forthright in his criticism of some London British Schools—and also provided mining with an exceptionally able and impartial investigator. Tremenheere is one of those Victorians who, faceless rather than eminent, amazes one both by his industry and versatility. He was instrumental in framing fourteen Acts of Parliament, sat as a commissioner on numerous committees, including important ones on education, mining and the employment of young persons, yet had the time to translate the works of Pindar into blank verse and write a book on the constitution of the United States. He spent a great deal of each year travelling the country, making on-the-spot investigations, and the width of vision gained from this, coupled with an essential thoroughness, made possible the development of a feature of the inspector's power which can be directly traced to Tremenheere's office. The inspector had to be a pragmatist who could not be intransigent, for the diversity of conditions in various collieries makes the enforcement of every section of the Regulations impossible [**doc. 4**], and this was what this first inspector came to see as he travelled between South Wales and Northumberland in the mid-nineteenth century. He saw that it was difficult to enforce the regulations relating to the employment of children in areas where the passages were so low that only boys were small enough to do the job. When someone, whom he describes as a 'philanthropic busybody', wrote to the Home Office complaining that Tremenheere would not enforce the sections of the Mines Act relating to the employment of children, he steadfastly refused to act, arguing that the practice would continue until those particular seams were worked out. This hard-headed realism, coupled with honesty, made him respected even amongst the tough, self-made colliery owners, and, because of this and his ability to see that it was no use promoting law unless you had the power of enforcement, he often achieved through persuasion more than he could have done through law court convictions. The ability to grant exemption under special circumstances, which dates from his time, continues to the present day and is provided for in a recent Mines and Quarries Act (1954).

Tremenheere should also be remembered as a tireless advocate of the comparative study of mining and mining law. For several years he spent his holidays touring the Continent, looking at French, Austrian, Belgian and German mines, corresponding with their managers and then reporting his findings to the Home Office. By showing that effective legislation had existed in France since their Mines Act (1810) and the Imperial Decree of 1813, he was able to stifle opposition from those who argued that interference with private property would not only be wrong, it would also be ineffective. When *laissez-faire* arguments were heard opposing the existence of mining inspectors he could show that the *Corps des Ingénieurs des Mines* was introducing safety regulations from 1813 onwards, and when other voices pointed out that a comprehensive code of safety regulations would serve no purpose he was able to show that a general code of thirty-two articles had existed in France for almost forty years. Tremenheere's pen was always active in support of improved mining legislation but he was a particularly able advocate of the desirability of strengthening the inspectorate and his is one of the hands which promoted the 1850 Act.

Yet in a sense this first inspector was very different from those who came after and whose power he advanced, for in certain respects he was more like a welfare officer than a modern industrial inspector. To an extent he was hamstrung by not being an engineer and because the 1842 Act did not contemplate underground inspection. The men who followed him were of a different calibre. First and

Coal Mining Law

foremost they had to be practical engineers, for the necessity of having seven years as a manager of a mine and passing an examination in mining science was written into the 1855 Inspection Act. Secondly, having a smaller area to report on, their local effectiveness could be greater. In 1850 there were four inspectors, in 1875, twenty-six inspectors and by 1906, thirty-eight inspectors, and this growth, together with their right to fine, made their powers real (**63**). At first they could only enter a mine on invitation since surprise visits were not authorised, or to investigate a fatal accident, yet by the end of the century spontaneous visiting was common and probably accounted for the bulk of the inspector's work. Yet, although the inspector's authority had increased and he had the power to inflict punishment, we should see that in the main his worth was as an educator. From their first publication in 1851 the Inspectors' Reports have not only listed casualties but have also given practical advice on the prevention of accidents [**doc. 11**]. The growth of an efficient inspectorate was, however, only one aspect of nineteenth-century legislation and important far-reaching acts were passed in 1855, 1860, 1872, and 1887.

Eventually in 1854 the possibility of formulating and legislating for a general code of safety was recognised in a Select Committee, so that when the 1855 Mines Act (18 & 19 Vict., cap. 108), which had to be introduced because the 1850 Act had expired, was presented it contained seven 'general rules' which applied to all mines and also required that individual collieries produced 'special rules' which had to be approved by the Home Office [**doc. 13**]. The act also recognised the principle of arbitration, but on the whole seemed to favour masters rather than men. This distinction between employee and employed is clearly felt when studying the punishments which could be inflicted for breaking the law. Men could be imprisoned for contravening the terms of the act but masters could only be fined. Trade unionists, like Alexander Macdonald, were not slow to point out the iniquity of this situation.

The 1855 Act was also a temporary measure, and in the natural course of events five years later the Government brought in the 1860 Mines Regulation and Inspection Bill (23 & 24 Vict., cap. 11). A lot had been learnt in the previous decade and therefore they felt justified in introducing this as permanent legislation. The age limit of underground workers was raised to twelve, unless the boy could show a certificate for reading and writing, and that of winding-engine boys to

The Industry

eighteen. The general and special rules were extended, and the powers of the inspectors increased, but probably the most far-reaching part of the act came in Section XXIX and related to check-weighing.

Payment by measure and the application of special rules about what was considered correct filling were, as shown above, a grievance which often caused strikes [**doc. 5**]. Corf sizes were not uniform, and, as the men pointed out, the coal trucks when replaced often 'growed a little'. This, together with criticism of 'special rules' concerning what constituted good coal and other problems involving pay rates, led to a lot of ill feeling. By 1860, however, the justice of the miners' case was generally recognised and when the temporary union of miners, formed to obtain 'a good inspection Act', pressed the point about weighing, the Government eventually agreed that the miners of individual collieries should be allowed to appoint from amongst themselves a checkweighman 'who should not interfere with the working but see and take account of the men's work'. The right to have a 'justice man', however, did not go unchallenged, and the famous Normansall case (**73**), in which a colliery disputed the appointment of a Barnsley trade union leader, is evidence of managements' reluctance to accept interference either from Government or men.

However, the significance of the introduction into the colliery yard of a workman who was responsible to and employed by the men goes beyond what was envisaged in the 1860 Act. The Government thought that by this innovation they were recognising the justice of the miners' case and removing an anomaly. In fact they were doing more than this, for by recognising the checkweighmen they were providing the men with leaders who could not easily be intimidated and thereby strengthening the infant trade unions. The checkweighman often combined the office with that of union lodge secretary, so that by the end of the century, as a result of this act, the national union became dominated by men who were severely practical, hard bargainers. To men like Pickard of Yorkshire, Abrahams of South Wales and most other mining leaders the initial step which led to high office was election to the post of checkweighman. When the debates took place in June 1860 on the terms of the act, it was the sections on the raising of the minimum age which attracted the most attention but in retrospect it is the checkweighman clauses, which slipped in by the back door, that have probably had the greatest effect on mining communities.

Coal Mining Law

The new Mines Act had only been in operation for two years when it had to be amended to satisfy the public outcry which immediately followed the Hartley Colliery disaster [**doc. 12**]. On 16 January 1862 the beam of the pumping engine at this remote Northumberland pit snapped, and fell into the shaft taking all the gear, lining and fitting. As a result of this and because there was no alternative route of escape, 204 men were buried alive. Petitions were sent to Parliament and eventually a proposal, which had been made by George Stephenson in 1835, became law. It became illegal to work single-shaft mines and the sinking of a second shaft was made compulsory. This amendment act, however, set the tone for the last great mining act of the century, the 1872 Coal Mines Regulation Act (35 & 36 Vict., cap. 76). There was still great concern about safety, particularly when it was shown that in the previous decade thirty-eight explosions had killed 1,650 men, and therefore the new act not only strengthened the inspectorate by agreeing to workmen's inspections, it also considered safety provision by increasing the number of general rules to thirty. Ventilation, timbering and lamps were reviewed and it was laid down that managers should have a certificate of competency [**doc. 14**]. A great improvement on any act passed up to that time, the 1872 measure ensured that the industry was forced to grapple with the surviving areas of inefficiency and danger.

From this time onwards legislation was realistic and practical. Following the Royal Commission in 1879 a Mines Act (50 & 51 Vict., cap. 58) was passed in 1887 which was concerned, amongst other things, with coal-dust explosions, shot firing and ambulance work. In a sense, because it considers the minutiae of colliery work, it seems an unimportant act, yet in another way it seems the natural culmination of fifty years of intense activity within Parliament and the unions. Three years before its passage eleven working men, five of them miners, had entered Parliament to join Thomas Burt, who had first represented a mining constituency in 1874. From then onwards because of the Parliamentary Reform Acts and the knowledgeable questioning of these men, who had, until recently, endured the disabilities legislated against in 1842 and the subsequent acts, Government found it difficult to ignore the complaints of miners. The national and local trade unions had by the turn of the century become powerful, and were exerting pressures which made sure that their views on legislation received attention.

5 The Coal Miners' Trade Unions

By inclination miners were conservative, and, although capable of class consciousness, they lacked political revolutionary fervour. Radicalism has rarely flourished in mining communities for concepts of solidarity did not easily extend beyond village boundaries. Remote from towns, insulated in 'single occupation' or agricultural villages, colliers seemed content to remain aloof from major reforming movements. Shoemakers were radicals, city artisans plotted revolution, but in the early nineteenth century the miner was a loyalist who, if given a chance, could usually be guaranteed to sing 'God Save the King'. Tory squires brought unenfranchised colliers to the hustings to intimidate political opponents, for eighteenth-century pitmen failed to see the relevance of a remote Parliament but enjoyed more than most the cursing, shouting and drinking. This swashbuckling attitude to politics partly accounts for their failure to take a leading part in the Chartist agitation. They were good trade unionists but poor politicians, for the central government seemed a long way off and unconcerned with their problems.

Miners only erupted into action when essential standards were threatened. On many occasions in the eighteenth century they revolted because of violently fluctuating bread prices. In these corn riots they stood beside other workers and by general consent were leaders in any fight. Physically tough, used to working in cellular units, they would sally out of their villages, attack the militia and retreat to the safety of the colliery workings. When roused they were difficult to suppress for they were capable of calculated violence—winding gins and corves were thrown down the shaft; the lives of merchants and overseers threatened. At Rhuddlan, North Wales, in 1740 miners sacked a granary, chased the dealer into hiding and threatened to 'cut his head off . . . and tie his guts about it'. The Newcastle Yeomanry were called out at the beginning of the nineteenth century after disaffected colliers broke open a wine cellar and left the resident viewer a note which complained of unfair social

The Coal Miners' Trade Unions

distinction and ended by stating that 'a great filosopher says to get noledge is to now wer ignorant'. Further south, Forest of Dean miners made the Mayor of Gloucester apprehensive when they stole flour, reduced its price and sold it in local towns. From this we see that the workmen who were fired on by troops in Yorkshire in 1893 [**doc. 17**] were part of a long violent tradition, which, though invariably lacking political overtones, was nevertheless based on a willingness to stand firm when intimidated.

There is little evidence of systematic combination for redress of grievance before 1800. In 1662, 2,000 miners in the North East signed a prepared petition to Charles II which complained of improper ventilation and bad treatment, but it was never sent. This suggests that although some collier organisation existed at that time it was isolated and probably ineffective. In the eighteenth century spontaneous public outbursts were common and expected; calculated long-term planning was not. There were some collier convictions following the passing of the Combination Acts (1799 and 1800), but usually pitmen preferred Friendly Society benefits to the more precarious transactions of illegal trade unions. Although not mutually exclusive—ambiguous clauses in Friendly Society rules often masked illegal union practices aimed at restraint of trade—the welfare benefits of the societies attracted the miner more than the radicalism of a militant trade union.

'The Friendly Associated Coal Miners within the township of Wakefield' met at the Three Tuns Inn and kept its records in a box with four different locks. The articles of this 1811 society allowed officers to have threepence each for liquor out of the fund. The officers 'feasted' on Easter Tuesday at a cost of eightpence for ale and eightpence for dinner but usually met to conduct business, collect subscriptions and pay out benefits. Each member paid an entrance fee of one shilling and a weekly subscription of threepence. When he had subscribed 10s 6d he could receive 6s 0d a week when injured or ill and be assisted if unemployed, except on grounds of misconduct (**73**).

Welfare schemes like this interested miners, but the sense of political discontent, epitomised by the march to St Peter's Fields in 1819 and the resultant massacre, only touched them lightly. In Newcastle in that year collier radicals met and deliberated, but the action of a large meeting of pitmen held in Leeds is more expressive of general collier attitudes. The West Riding men did not come

The Industry

together to support manhood suffrage but because a demand to sign a contract document offended a sense of 'privelege inherent in Englishmen'. Three thousand paraded to the accompaniment of brass bands, sang the National Anthem and 'Rule, Britannia!' before they peacefully dispersed. It was the same in the North East for in this year of turbulence, although a few colliers marched, the vast majority remained at work. The Northumberland pitmen had struck in 1810 [**doc. 8**] against an alteration in the terms of the yearly binding, but from that date until 1830 their unions remained relatively peaceful. That year saw the formation of 'Hepburn's Union' and in the following April there was an attempt to amalgamate with other unions to form a loose federation. Delegates joined with representative miners from Staffordshire, Yorkshire, Cheshire and Wales and, giving voice to the demands of 9,000 men, sought membership of Doherty's National Association of United Trades. Permanent district mining unions began to appear and pursue limited trade objectives but the political questions raised in the manifestoes of small artisan unions are absent when colliers stated their grievances. The North Eastern worker in 1831 objected to 'the Bond', laying off, eviction for a petty infringement of colliery rules, child labour and 'Tommy Shops', but did not seek redress through parliamentary action. There was a demand for equality but it was expressed in a primitive form, as when in 1792 Sunderland colliers accosted Lord Lambton and said, 'We like his [Tom Paine's] work much, you have a great estate, General; we shall soon divide it amongst us.' They lacked a formulated, decisive political creed. Just as the conservative union leader Pickard, at the end of the nineteenth century, dismissed young leaders who preached socialism as men 'who wish to build ladders to the moon', so did the earlier rank and file regard the teachings of Robert Owen. They were unmoved by the growth of the Grand National Consolidated Trade Union (1834) and also by its fall. Strong and often rich enough to stay aloof they did not identify with less fortunate workers.

Of course in many ways they were better off. Engels describes their standard of living as fairly good and their wages high—except in parts of Scotland and Ireland—in comparison with surrounding workers. Their working conditions were extremely dangerous and harsh but otherwise they did not suffer the privations experienced by labourers in manufacturing towns. 'Collier lads get gold and silver; factory lads get nowt but brass' is a colloquial way of ex-

The Coal Miners' Trade Unions

pressing a truism, for, although methods of collective payment make it difficult at this early date to produce an accurate wage index, there is little doubt that miners were well paid. Arduous working conditions, remoteness and relatively high wages made them men apart. To this must be added the fact that many were not wage labourers in the strict sense but sub-contractors or partners in 'butty' gangs. This meant that until the middle of the century when mines grew into major industrial units, the struggle of employee and employed was not in clear perspective. Rugged individualism survived long after other workers had sought strength in collective action. Mining unions, such as they were before the 1830s, were impermanent organisations which arose to fight for limited objectives and then disappeared, often leaving the leaders destitute and blacklisted.

Then in the 1840s militant trade unionism came to the coalfields. The year 1844 saw the outbreak of a prolonged strike in Northumberland and Durham, several lesser strikes in other areas and attempted support from a newly formed national union (75). This union, the Miners' Association of Great Britain and Northern Ireland, probably began when, in November 1842, the secretary of the Miners' Philanthropic Society in Wakefield asked colliers in other districts to communicate matters of interest to miners for inclusion in a periodical. Using Chartist phraseology, speaking of owners as 'the greatest tyrants on earth', they commented upon and solicited information about numbers employed, accidents, wage reductions and strikes. The initial response to the Yorkshire appeal was disappointing. Having avoided activity in Chartist outbreaks, many colliers seemed reluctant to correspond or to see the solution to labour grievance in trade union action on a national scale. Then slowly the summons was heeded. Three North Eastern collieries answered and from that time numbers grew. The columns of the Chartist *Northern Star* were used, and by May 1843 a delegate meeting in Newcastle was strong enough to support missionaries in areas which had not joined the society. Advocating organisation similar to that used by the Grand National Consolidated Union, they taught that British colliers could be welded into a coherent body if they formed colliery societies which sent delegates to district committees. 'One grand body' of district delegates could then be formed to establish an effective strike fund and, having national authority, direct general policy by demanding a measure of uniformity

The Industry

throughout the country. As membership grew—there were 50,000 members by autumn 1843 when 200 delegates attended the first National Conference—inevitably the leadership passed from Yorkshire to the numerically strong North East. David Swallow, the Yorkshire Secretary, was replaced as leader by Martin Jude, and all the chief officers were Durham or Northumberland men.

Not all the leaders of the movement, however, were practising miners. William Bressley was an Accrington chairmaker, William Daniels had woven carpets in Edinburgh, while the most influential recruit to the miner's cause was a Bath attorney, W. P. Roberts, who had devoted a lifetime of travel and hard work to advancing the claims of the miners. He showed that the courts, even when they had coal-owner J.P.s and could not be guaranteed to be completely impartial, were useful as a public platform. A recruit from the Chartist movement, he fought on behalf of the men in the courtrooms of the North East and South Wales, and was particularly popular in Lancashire. His melodramatic oratory was in tune with the times, and if a later generation of union leaders despised him because he swayed an audience with emotive language, he was deeply loved in the pit villages. 'Stick unto your Union and mind what Roberts say' (**40**) was a ballad-monger's expression of this affection which could not be destroyed by accusations of his working for personal gain. This influential Chartist lawyer stood firm behind Martin Jude when the 1844 strike broke out.

The Northumberland and Durham miners ceased work as planned on 5 April, and from the first the owners showed determination to fight with every available weapon (**70**). The horses were brought up from underground—always a clear sign that there was little hope of an early settlement—and the management prepared to starve the men into submission. The miners for their part adopted a policy which stressed solidarity and the maintenance of good order. From the first they solicited public sympathy by presenting their case in the columns of newspapers and at large well-attended meetings in Durham and Newcastle. These gatherings were fully reported in the miners' own periodical *The Miner's Advocate* [**doc. 16**] and, with some bias, in the local press. The colliers suggested that 'the masters set us an example, for the masters formed a union for the protection of their interests' which had introduced stringent bonds and a material reduction in wages. Public sympathy was forthcoming for although the local press, which was probably subject to ownership

The Coal Miners' Trade Unions

pressures, was critical some of the London papers, the new satirical paper *Punch* for instance, gave support and encouragement to the men (**70**).

In the second month of the strike, probably because there were no signs of union weakness, the owners' attitudes hardened. Agents were sent south and blackleg miners brought in from Wales and Cornwall. Sometimes these colliers came innocently believing that new collieries were being opened. If they did they were reasoned with and usually accepted the union's return railfare. Others did not go so easily and these men and their families reaped a bitter harvest when eventually the recalcitrant men were forced back to work [**doc. 6**]. To provide accommodation for the blackleg labour, eviction of striker families had to take place, and because of this direct intimidation shanty towns, built of beds and chests of drawers roofed with bedding, huddled to the fringes of colliery villages throughout the long hot summer. While the weather was reasonable good humour persisted. Clocks, wedding rings and other valued possessions were pawned to buy food and in adversity communal identity was strong. Even the powerful Marquis of Londonderry's attempt to starve the men by threatening local food suppliers did not goad the strikers into violence. Hepburn had told them in 1831 that, 'To know how to wait is the secret of success', and, knowing that it was the object of the owners' strategy to make them break the peace, they remained calm, debated, passed resolutions but did not riot.

Miners in several other coalfields, having first given financial support, now gave active encouragement to the North Eastern men and also struck. On 12 May the Yorkshire, Derbyshire and Nottinghamshire men stopped work. In Yorkshire the demands of the men were not uniform nor was their enthusiasm. In some districts few colliers responded to the strike call, in others men stayed out for twenty-four weeks and did not return to work until a month after the strike had ended in the North East. Hardened unionists from other trades supported these activists. The North was experiencing Chartism and in Sheffield the small radical groups gave support. The white-metal smiths and the table knife forgers organised processions and delivered bread to the strikers as they recognised the political overtones which accompany strikes by vast bodies of men. Yet, despite this and the tremendous sense of injustice, the strike was doomed and this became more apparent as summer passed into autumn. Resolution weakened as the weather changed and what

The Industry

seemed tolerable living conditions on a close, hot night became terrible as the camping sites were clogged with mud. A meeting, held on Town Moor, Newcastle, on 30 July and attended by a crowd of 30,000, took place in pouring rain. The money expected from a sympathetic public only trickled in and union agents reported the hopelessness of collecting subscriptions. Men left to find work in other areas and rumours circulated saying that the union had collapsed. The restraint which had been a keynote of the early weeks was lost. Blacklegs were attacked and riots took place. Confronted by starving wives and children, the miners drifted back to work so that when another Town Moor rally was held on 13 August, only 12,000 miners came. In the end arbitrary and immense wealth proved stronger than trade unionism.

Some positive improvement did, however, take place following this strike. Contract by yearly bond, one of the grievances which precipitated strike action, disappeared from the Northumberland field. It lingered on for thirty more years in Durham, but even there by 1872 fortnightly hiring was practised in most districts. Other legacies of the strike were not so pleasant or beneficial. The Miners' Association, for instance, was broken, its leaders dispensed and its members forced to return to the old wage rates. The tolerance and peace likewise ended and the blackleg miners, who during the strike had protection from the police, found themselves alone in hostile communities. A colliery gallery offers ample opportunity for ambush and victimisation. There the blacklegs' children suffered indignities at the hands of vengeful colliers; they were tipped from trucks by ropes stretched between pit props and left in the dark. The blacklegs themselves suffered worse, for they not only lost self-respect; they also lost money. At that time wage payment was based on output and it was the practice to hang an individual's tally disc on full corves before they went to the pitbank. When the strike was over it became common usage for the local men to remove these markings from blackleg corves so that when the interlopers surfaced, having put up with abuse and violence below ground, they sometimes found that there was little or no financial compensation. Although the great strike was over its lessons lingered and its effects echoed in the coalfield.

After the 1844 collapse the ideal of a national union was discredited. The county unions weakened, until by the end of the decade they had virtually disappeared; their leaders were victim-

The Coal Miners' Trade Unions

ised and some had left mining altogether. David Swallow 'entered into other pursuits and settled down'; 'The Twelve Apostles' who travelled to London to state the miners' case were sacked: Edward Richardson, an independent North-Eastern leader, after selling the books 'he prized so much, died in South Shields from sheer want' (**70**). Fourteen years passed before an effective county union—the Yorkshire Miners Association (1858)—reappeared (**68**). This does not mean that trade unionism was dead but only that it was changing its form, and instead of putting faith in the emergence of a national organisation it was adopting limited objectives and strengthening itself at lodge level. Conflicting opinions have been expressed about what happened in the years immediately following 1844. In a well-known report quoted by the Webbs and others (**66**), Alexander Macdonald, speaking from personal knowledge to the 1873 Leeds Conference, suggests that at all levels unions withered to nothing: 'At the close of 1855, it might be said that union among the miners in the whole country had almost died out, the fragments of union that existed got less by degrees and more minute.' G. D. H. Cole, on the other hand, writing in 1937, believed that too much was read into this passage. He considered that all that happened was that local unions just ceased to act together following the collapse of Martin Jude's National Association. The truth probably lies somewhere between and close to an opinion expressed by John Holmes, a Leeds businessman sympathetic to the miners' interest. Writing in 1863, he said that unions arose to settle immediate disputes and that their organisation did not extend beyond the colliery lodge. These colliery units, however, were often quite strong and, if loosely linked with neighbouring lodges, could engage management with a sustained strike. Such cells produced the trained leaders who took control of the big unions at the turn of the century. Unionists like Ben Pickard, who kept the books for his father's Kippax lodge in his twelfth year, and Thomas Burt gained their early experience of industrial organisation in their colliery's union lodge. The checkweighman clauses in the 1860 and subsequent acts further strengthened this unit by providing the men with their own officials. From that time onwards, each colliery had a permanent full-time negotiator and trade unionism developed. That the appointment of such men was of unmitigated benefit is, however, open to question. It is true a checkweighman appointed by the men was not easily victimised. Fair and accurate weighing resulted but with it came profes-

The Industry

sionalism (**76**). The union leader became a full-time official who was set apart from his followers because his job was different. The men who were appointed to this post usually had characteristics in common. They were sober, respected and businesslike but because their livelihood depended on the pit remaining open they were also under pressure to conform. In 1860 the law said that the checkweighman had to be appointed from among a colliery's employees, and, although management could not dismiss an official after he had been appointed by the men, it was not compelled to re-employ him after a strike or lockout. It was therefore judicious to counsel the prudence which avoided strikes, and increasingly the checkweighmen union leaders became immersed in the details of negotiation, procedure and welfare schemes. Judgement by neutrals seemed the panacea of industrial peace and they actively supported the formation of Boards of Arbitration and Conciliation, even after it was recognised that rulings rarely favoured the men. Such moderation definitely allowed the unions to have sound financial and administrative foundations, but daily contact with management led to overt conciliatory action. The fact that checkweighmen sat as equals to negotiate sliding-scale agreements sometimes blunted their vigour in discussion and made them oppose the wishes of the men they represented. They lectured both reluctant owners and recalcitrant men. This experience made them suspicious of grandiose plans for a militant national union. Instead they trusted their knowledge of local conditions and this allowed them to become the architects of new county unions.

Yorkshire colliers have had permanent union officials ever since the South Yorkshire Miners' Association was formed in 1858. In the 1860s other county unions appeared. The Northumberland and Durham Miners Mutual Confident Association was formed in 1863, when pitmen agreed to pay a fortnightly contribution of a penny so that they could receive welfare benefits and support victimised miners. It is significant that, although at this time each lodge retained its own funds, it agreed not to undertake strikes without the approval of the general management committee. In this there seems to be an attempt at a compromise which sought to recognise central union authority and also to support the growth of independent lodges. This reflects a restraint which sees the importance of building on firm foundations. Many believed that the earlier unions had floundered because they asked far too much too quickly; all came

The Coal Miners' Trade Unions

out together and became involved in long unsuccessful strikes in which welfare moneys were eaten up. Union members saw the importance of strong but small units and the miners' lodges were the strength points of the late nineteenth-century miners' unions. Democratic, usually based on a single pit, meeting regularly under elected officials, using agendas and systematic business processes, these were the units out of which grew the district, county and later the coalfield unions. In the fact that these larger unions were associations of collieries before they were associations of men lay a great source of power.

The call for nation-wide union organisation first came in letters to John Tower's *British Miner* but its cause was soon taken up by Alexander Macdonald. Called by Karl Marx 'miserable Macdonald' yet seen by the Webbs as the man to whom 'the miners owe their position in the world', this Scottish miner continues to be a controversial figure. He was born in Lanarkshire in 1821, started work as a pitboy when eight years old and continued as a miner for the next sixteen years. Up to this time his background closely resembles that of the other young men who in later life were to become mining leaders. His subsequent history is, however, more unusual. Having taught himself Latin and Greek at evening classes, he went to Glasgow University, paid class and lodging fees out of what he had earned in the summer months, and eventually took a degree. For a time he managed a mine, taught, indulged in financial speculation but eventually decided to devote himself to trade union activity. Standing midway between the casual amateurs who had, under tremendous pressure, led the earlier unions, and the highly organised bureaucratic mining leaders of the turn of the century, he represents a new brand of trade unionist. When he assumed office, controversy continued as to the best method of securing benefits for miners. One school, the natural heirs of the Miners' Association and containing many veterans of that union's fight, still favoured an uncompromising policy standing firmly against management and accepting the reality of class war. It was a policy which if successful could bring with it great gains, but the costs were high for such attitudes would involve strikes and demanded unity, large membership and high subscription (**76**). Another group sought to play a canny game. They believed in moderation, arbitration and the promotion of class harmony; because this cost so much less to sponsor and the necessity of having unity need not be placed at such a high

The Industry

premium, it appealed to many miners. Macdonald soon achieved leadership of the latter group, put his faith in Parliament and in the advocacy of mining acts to secure true weighing, education of the young, trained management and improved safety regulations.

In 1858 he convened a meeting at Ashton-under-Lyne which was attended by delegates representing 4,000 men. The success of this meeting brought about the formation of the National Association of Coal, Lime and Ironstone Miners of Great Britain, better known as the National Miners' Union (**79**). The professionalism of Macdonald's approach is immediately realised if the first major conference in Leeds in 1863 is compared with what went before. Earlier meetings had been formless interchanges of opinion, but from the outset, when it adopted procedure modelled on the recommendations of the National Association for the Promotion of Social Sciences, this union gathering set a pattern for conference efficiency. What it could not do, however, was to achieve unity. It was agreed by many of the delegates, some of whom seemed self-appointed, that they would unite to promote legislation but would not have a coordinated policy for industrial action or collective bargaining. Arbitration was favoured but centralisation of funds was opposed. It is difficult to tell if this was the policy favoured by the majority of the rank and file. Genuine widespread support came from Macdonaldite strongholds in Yorkshire, but from other fields, particularly Scottish ones, membership was either apathetic or, as in the case of Lancashire, openly hostile to Macdonald and his group. Dislike of the policies which he advocated also came from South Wales (**71**), for in that area the old Chartist advocate Roberts retained a popular following and Macdonald had no love for the policies which the lawyer continued to forward. Roberts denounced the coal-owners in forthright language, advocated strikes and with the melodrama of a tested attorney appealed over the heads of the delegates to popular opinion. In doing this he used all the devices which Macdonald, ever conscious of the need of a respectable image, feared and hated.

In 1864 the gap between the two policies—moderation or militancy—became most real when a group calling themselves the Practical Miners' Association broke away from the parent body. This union soon disbanded, broken by a disastrous strike in South Staffordshire, but in 1869 a union called the Amalgamated Association of Miners was formed and for a few years became Macdonald's

The Coal Miners' Trade Unions

opposition. Like the Practical Miners' Association this body favoured centralised control, taught that the National Miners' Union would not succeed in organising weaker districts and said that until there was wage standardisation blackleg miners could always be introduced to break strikes. Macdonald's supporters wanted strong independent local association but the new union, led by Thomas Halliday, wanted national systematic support for local strikes. The debate divided the colliers but increased union membership. In 1873 the two bodies were well matched. Meeting in Leeds, the National Miners' Union claimed the support of 123,000 miners, and at a conference in Bristol the Amalgamated Association said that it represented 99,000 colliers (**79**). Scotland, the North East and Yorkshire favoured Macdonald's policy, but Lancashire and Wales stood behind Halliday. Some areas, the Midland field for example, present a picture of divided loyalty. The 3,900 miners of Dudley supported the National Union but six miles away in West Bromwich the workers were Amalgamated Union men, However, this rivalry did not last long, for in 1875 the militant union was wrecked by the usual cause, a prolonged lockout.

In contrasting these two unions we see trade union perspectives which differ fundamentally from those usually presented. Until recently the pacific tendencies of unions in the period 1850–75 have been over-emphasised (**79**). The centralised craft unions—the Amalgamated Society of Engineers and the Amalgamated Society of Carpenters and Joiners—which, with concentrated energy and business acumen, fought for parliamentary legislation, have been thought characteristic unions. We now know that they were not representative of the entire movement and although moderate, peaceful policies could, in the case of the carpenters' union, be linked with centralisation, in the coal unions the reverse happened. For it was the militant miners in the Amalgamated Union who clamoured for a strong centralised authority and the peaceful, moderate Miners' Association which favoured decentralisation, arbitration and the parliamentary lobby. The attention which the National Miners' Union paid to legislative advance brought dividends and major reforms affecting mining operations were achieved because of its interest, but it did not pursue labour and wage questions with the same determination. It is difficult therefore to assess the extent to which it represented grass-root aspiration. The accusation that it was but a natural heir to the traditions of the 'Pitmen's Grave and Coffin

The Industry

Clubs' was heard for it seemed too concerned with safety and not sufficiently interested in wages. If this accusation was true, then the National Union was only fulfilling the resolution passed at its inception which remained the union's guiding principle until the union withered and was replaced by the Miners' Federation. The clause passed by the 1863 conference defined the purpose of the National Union and stated that, although the officers should be concerned with 'legislation for the inspection of mines, and for compensation for accidents, . . . they were not expected to interfere with any local disputes'.

This then was the governing principle of the union which in 1875 absorbed the last remnants of the broken Amalgamated Association. For a short time Halliday became secretary of the surviving body, but in the hard thirteen years from 1875 to 1888 confidence in this remaining national union weakened. There was general approval for pressure to secure better law but without stimulation from a militant group the general membership became disenchanted and eventually apathetic. Numbers dwindled so that in Yorkshire, in more prosperous times a bastion of Macdonaldite strength, in a labour force of 60,000 men only 2,800 belonged to the union. In areas other than this county and the North East matters were worse, for eventually the miners came to believe, as they were forced to accept large wage reductions, that conservation of district strength was the only possible policy for survival.

Then gradually they became disillusioned with this formula for by 1880 there no longer seemed security in the formation of district units. An extensive railway network which, for instance, allowed coal from inland coalfields to compete with Newcastle sea-borne coal delivered to the upper Thames valley, was militating against local combinations of any kind (52), be they unions of men or masters. Since transport costs were no longer decisive, coal prices and wages could only be controlled if output was restricted and, since this had no validity if confined to one area, the emergence of a national union became a necessity. The history of mining conferences had always been a history of attempts to limit production, but eventually in 1889 a combination of favourable circumstances made this a possibility and a strong national union was formed which secured considerable advances.

By the 1880s sliding scales were discredited. Where they survived they did so because the unions were weak and could not resist (77).

The Coal Miners' Trade Unions

In general the men believed that sliding scales always operated against their best interest, for they encouraged management to sell at a low price when trade was depressed and to meet this drop in profit by wage reduction. The demand for a minimum wage and for limitation of hours was therefore heard. This was accompanied by recognition that a wage reduction in one coalfield snowballed and affected another region. Realisation that coal could be quickly brought into strike areas and thereby partly nullify action made many unionists see the necessity of coming to some sort of agreement with miners in other districts. Such an agreement was achieved in 1883 when 8,000 North Staffordshire workers struck rather than accept a 10 per cent reduction, and Lancashire unionists came to their aid and levied ninepence to give them a guaranteed ten shillings a week. Thomas Ashton, writing at the time, asked, 'Who amongst our ranks would rather give the employers two shillings than the men eightpence? This and more you will have to do if the battle is lost.' On this occasion even non-union men subscribed when they realised that a North Staffordshire reduction was a Lancashire problem. In the end, however, it was Ben Pickard of Yorkshire and Thomas Ashton of Lancashire who convinced the bulk of the men that in the final analysis they had common problems.

Ben Pickard was one of the great mining leaders of the nineteenth century. Stocky, brusque, something of a bully, this Yorkshire leader's common sense appealed to the miners and he enjoyed enormous prestige. He had been a union official since he was twelve years old and as a young man became a paid secretary of the West Yorkshire miners (**73**). During the 1870s he had supported Macdonald and came to be vice-president of the M.N.U. In 1886 when Pickard entered into a full concordat with Ashton he was the secretary of the Yorkshire Miners' Association, a union which he had directed after securing the amalgamation of the county's two district bodies. Thomas Ashton, on the other hand, had been a member of the Amalgamated Association and represented the militant element in the partnership. Pickard and Ashton were well matched, tireless workers and if, as orators, they could not compete with Macdonald or Keir Hardie, at least they had exceptional executive skill.

Negotiation between these two secretaries began informally in 1881 after Ashton had succeeded in bringing several small local unions together in the Lancashire and Cheshire Federation. Both were officials of broken, weak organisations, but because neither

The Industry

organisation was from an area with large exporting markets, subject to dramatic price fluctuations, their union slowly became quite real. This was movement towards wider co-operation and from this trans-Pennine link there grew something much bigger. Pickard and Ashton urged the formation of a union which was more than a loosely knit federation. They wanted a minimum wage policy and an eight-hour day. Gradually other unionists listened and despite membership losses during the 1885–7 depression union strength grew as prices rose in the two years which followed this recession (**65**).

At the Edinburgh Miners' Conference in 1887 the point had been reached at which federation could be realistically debated. The opposition to the idea forwarded the anticipated arguments as Thomas Burt pointed out that central authorities needed large funds, and many unionists would not accept advice on when to initiate action against employers. Yet without federation the other important item on the agenda, one seeking to limit output, could not be achieved. On this proposal too there was a great divergence of opinion. One group thought that parliamentary action for an eight-hour day (**78**) was the solution, while another section urged that the men should be advised to stand idle for a part of each fortnight. Both were legitimate methods of controlling output and consequently wages, but the delegates could not be sure which would be the most effective method to adopt. Edward Cowey of Yorkshire pointed to this persisting predicament when he said 'You first pass a resolution to do it yourselves and then in the next breath you say to them, "Be quiet, the Government's going to do it for you" . . . Choose ye this day whom ye will serve.' In 1889 a great body decided that they would support the leaders who counselled industrial action, federate and make federation a reality.

In 1889 the time was ripe for federation. Three years earlier in Nottingham a speaker had indicated that the best way to obtain a 15 per cent increase in wages was 'by laying the whole community idle and subsequently limiting output by less hours or less days'. This was said when unionism was weak, but at the end of the decade it was growing in strength. Coal prices were also rising and in the autumn of 1888 began the first successful wages movement for many years. In October 1888 a meeting of delegates from areas outside the North East, and free from sliding-scale agreements, was held at Derby. The South Wales men, who were still governed by the scale which they had been forced to accept following the disastrous 1875

The Coal Miners' Trade Unions

strike, were absent but delegates from Lancashire, Derbyshire, Nottinghamshire, Leicestershire, Stirlingshire and the powerful Yorkshire union were present (**68**). They learnt that an earlier demand for 10 per cent increase had been met in most districts but that in Yorkshire 30,000 men were on strike. On hearing this the conference resolved to stand firm, promote a shilling strike levy and wait for the owners to react. When eventually the employers met in Sheffield they capitulated and unconditionally granted the increase. Unable to achieve even the semblance of unity and because prices were rising, the owners agreed to what seemed just another wage increase but what they almost unknowingly did was to underline the value of a miners' federation by showing that a national union could secure more than mining laws.

The Miners' National Union met in January 1889 for the last conference of a body which had, first under Macdonald and later under the influence of Burt, achieved so much effective parliamentary legislation. Now its tone was out of tune with the times for, although formal federation did not exist, the men, having seen that national unity could produce pressures for an effective wage advance and knowing that the fight for major legislative action was won, called for more industrial action. The leaders from the North East were slow to accept this change of spirit. They tried to broaden the base of the old union, and when the most important conference of that year was held in March in Birmingham, Northumberland and Durham leaders were absent. The number of delegates who did attend, however, represented a wider cross-section of the mining community than had attended the 1888 Derby meeting. Delegates representing 317,515 men were there and in a spirit of unity bred from success they called for another 10 per cent rise in pay. Again the owners gave in and the increase conceded in two instalments was paid by the late summer of 1889. When the Somerset owners resisted, the union sent the treasurer, Enoch 'Sammy wi' the bag' Edwards, to pay out £10,000 in strike money and the South-Western men got their pay. Still unsatisfied the men yet again asked for 10 per cent and saw it granted. Within a year the men had demanded and received substantial pay increases. In Stirlingshire, where the average price of pithead coal had risen 71 per cent between 1887 and February 1890, the pitmen got 55 per cent more in wages, and, although this was exceptional, most of the miners in the loose-knit federation were 30 per cent better off. The rises suggested

The Industry

that the dividends of centralisation were handsome and so it was natural that the men should seek to rationalise the situation and formally bind themselves into a union which recognised their new mood. At Newport the *ad hoc* group which had first met at Pickard's invitation in 1888 formed the Miners' Federation of Great Britain and by so doing created the biggest union in Europe (**68**). Unlike the earlier unions it was very powerful, and if the North-Eastern leaders continued to believe that improvement could be best achieved through parliamentary lobbying, by 1890 a majority of mining unionists favoured the exertion of trade pressure.

The other important development of this period was the appearance of an international miners' conference. John Wilson, the Durham leader, says that the idea originated when Thomas Burt and Charles Fenwick attended a Possibilist conference in Paris and there received a letter from a Marxist conference, which was meeting in the city at the same time, asking for an interchange of opinion. Subsequently eighteen delegates met at a dingy coffee house and Great Britain was asked to initiate an international conference. Consequently at the M.N.U. meeting in 1889 J. Toyne proposed that 'the conference accepted that an international working miners' conference has become a necessity' and pointed out the identification of interest which existed between the English miner and his continental counterpart. He stressed that the foreign collier would learn to recognise the benefits of powerful trade union organisation and the English worker would not be intimidated by the threat that cheap labour would under-cut coal prices when both groups were members of an international conference. Although the proposal met with general approval there was spirited opposition from Pickard, who mistrusted the type of congress 'at which certain gentlemen living in London because they belong to some federated body, should speak for the French, German and Italian miners'. He thought that it was impossible to hold a valuable conference of this type because foreign workmen would not be able to attend. Whether Pickard's reasoning was influenced by the fact that North-Eastern leaders sponsored the motion it is difficult to say—he always bristled when John Wilson spoke—but despite this opposition from the Chair it was resolved that an International Miners' Conference be convened, and in 1890 this met for its first session at Jolimont in Belgium. Immediately the power of the British miners was recognised as their delegates joined miners from the enfeebled Belgian, Bohemian and French trade unions. In France, despite its early history of mining legislation, women were still employed underground and men worked thirteen-hour shifts. Legislative battles which the British miner had fought and won forty years before still concerned the foreign worker. In such circumstances the call for the eight-hour day seemed unrealistic but it was proposed and adopted as policy. However, when a call for a general strike in support of this motion was proposed, Hardie got support from the continental members, but the other British leaders, ever cautious, deferred a decision, and in what was thought to be a way of shelving an embarrassing question, proposed that they take the issue back to the men. It was thought that although the British miner would support the continental worker's fight financially he would not strike on his behalf. This opinion proved wrong when two out of every three British miners agreed to support such a strike. It was not lack of enthusiasm from this side of the Channel which made sure that such international action did not take place, but lack of resolution on the part of French and German delegates. The British leadership, having breathed a sigh of relief, gave financial support to Belgian miners who had already embarked on a costly, and as it proved abortive, strike but from then onwards, although the International Conference continued to meet annually until 1914, it had lost its power. Discussion of industrial problems and welfare schemes took place, peace and fraternal loyalty resolutions were passed, but the will to take positive militant action was absent.

The support of the proposed international miners' strike and the leadership's failure to comprehend that rank and file were ready to commit themselves on wider issues typified what was happening in the mining unions at all levels and in all areas. By 1890 it was becoming apparent that the leaders, who twenty years before had built up the county associations and made these unions effective, were out of touch with the certain realities within the new situation. They were underestimating the aspirations of many of their young members. The Durham and Northumberland unions kept out of the M.F.G.B., more because their leaders disliked the idea of centralised control and could exploit regional working variations to frustrate amalgamations than because this was the opinion of the bulk of the membership. When forced into the unpalatable position of having to join the M.F.G.B. in 1892, Wilson's heart was never with its policies and he used his power to sabotage the large union so that Durham withdrew a year later. Similarly in South Wales 'Mabon'

The Industry

Abrahams, having chaired the Joint Sliding Scale Association, could not see that many of his South Wales men regarded such industrial relations machinery as a 'starving scale' and demanded instead a minimum wage policy. Pickard might support a middle-class Liberal in a Barnsley election and speak against the Independent Labour candidate but he could not stop the younger worker looking beyond an 'eight-bob-a-day' policy and seeking his salvation in socialism. After the 1893 strike, with its attendant misery and death [**doc. 17**], new leaders preaching class-war doctrines and advocating far-reaching aggressive strikes received an audience. For many the period of political caution was at an end.

Part Three

THE CONSEQUENCES

6 The Aftermath

The coal industry has undergone such fundamental changes in the last seventy years that it is impossible concisely to chart its fortunes. Since this final chapter can only present some important aspects of twentieth-century mining it might be best to look first at the state of the industry immediately before the first world war and compare this period of impressive achievement with subsequent development.

In the years 1913 and 1914 the British coal industry produced its record 287 million tons, the Miners' Federation became the biggest trade union in Europe and a million miners struck in a coal dispute of unprecedented magnitude. At that time almost every aspect of the industry was grand and spectacular. It was the period when progressive owners ousted inefficient small colliery management in favour of large powerful district amalgamations. The biggest seemed best and even the world's shipping depended on the 32 per cent of the nation's coal output which was exported to foreign bunkers. Inside Parliament the influence of the miners' representatives increased, for as the Labour Party grew so too did the power of the M.F.G.B. Of the forty-two Labour M.P.s elected in December 1909 eighteen were miners. Since affiliation in the previous year the union fund sustained the young party and when the electorate gave more support to Labour candidates mining communities became pocket boroughs for M.F.G.B. nominees. The miners' lobby quickly became the best organised pressure group in Parliament and the Triple Alliance, which the union leaders initiated in 1913–14 with railwaymen and transport workers, was the strongest union association in Europe. Conscious of its strength the miners' union adopted far-reaching and adventurous policies. The Socialist Robert Smillie had replaced the cautious Pickard as president of the union, and nationalisation of the mines, first accepted in principle in 1894, was urged by his executive and presented as a parliamentary bill. From the 'left' other strident voices were heard. The much-quoted pamphlet *The Miners' Next Step* urged that 'the old policy of identity of interest

The Consequences

between employers and ourselves be abolished and a policy of open hostility installed'. This was not the mainspring of the disturbances which sent the army to Tonypandy in 1910 and precipitated the national strike in 1911–12, but it was symptomatic of a general desire for change. The industry was conscious of its prime importance and both employer and employed were aware of their power. Even that ghastly indicator of pit size, the major colliery disaster, symbolically confirmed that the industry flourished by recording in 1913 an unparalleled accident at Senghenydd in which 439 men were killed. When the Expeditionary Force left for France in 1914 Britain still believed that her economic prosperity rested on a bedrock of coal. Thirty years later when another British army returned from Europe this was not the case.

The change which took place in the period between the wars was partly caused by fundamental alterations in the world economy which resulted in a drop in the demand for British exported coal. During the 1914–18 war a depleted labour force and restricted export facilities masked the fact that many countries no longer needed British fuel. For a time this reality was disguised by events which supported the illusion that Britain still had a stable overseas market. The 1921 strike, the American strike a year later and the French occupation of the Ruhr created freak market conditions and it was not until there was a sharp fall in demand in 1925 that the industry became aware that circumstances had changed. Fifty million tons were exported in that year—a figure which compared unfavourably with the 73 million tons exported in 1913—and from then onwards there was a constant downward trend. Expansion of the domestic market was slight and did not make up for this drop in demand. The fuel requirements of the electricity and gas industries barely compensated for the decreased requirements of iron and steel founders and the challenge of other fuels, particularly oil, and greater fuel efficiency, resulted in a steady decline.

Attempts to arrest this process were unsuccessful. Both Parliament and the private sector intervened. Amalgamation and efficiency became the keynote of the late twenties and thirties. Sir Charles Markham's statement to the 1924–5 Royal Commission, that small pits want 'shutting up . . . To-day a colliery is no good unless it raises 20,000 tons a day', reflected the attitude of advanced management. Britain had been slow to adopt mechanised mining and in 1900 only 2 per cent of the coal was mechanically cut but at the

The Aftermath

outbreak of the Second World War 54 per cent of the coal was cut in this way. Parliament also tried to restore the industry's fortune. The Sankey Commission (1919) advocated nationalisation and that headed by Herbert Samuel encouraged further amalgamation through recommendations embodied in the 1925 Mining Act. Yet such courses could not turn back the clock and eventually the coal industry accepted that its place in the national economy had altered. It was no longer the pre-eminent source of power for the challenge of oil and electricity was permanent.

In 1946 the Coal Industry Nationalisation Bill was passed and national planning made it possible for the radical remedies to be implemented. The National Coal Board became the biggest employer in the country with a responsibility for nearly 800,000 workers and an annual turnover of £360 million but it still faced many problems. The 'People's Pits' did not herald the millennium but at least they made it possible to rationalise production. Unfortunately for many this meant pit closure and as many collieries in the North East and South Wales were closed the miners were made redundant. Despite the fact that modern pits continued to be sunk along the eastern edges of the Yorkshire and Nottingham field continuing contraction was accepted by both management and men.

Acceptance, however, did not mean an end to frustration for strikes like the 1969 Yorkshire stoppage, which on the surface appeared to be about wage rates, were probably an expression of the hopelessness which colliers feel as they see the local pits closing and know that there is a terrible inevitability in the process.

It is difficult to find a document which adequately captures the prevailing atmosphere of mining communities. The musical play *Close the Coalhouse Door* managed to give insight into modern colliery village life as did sociological works and novels like Coombes' *These Poor Hands*, Orwell's *Road to Wigan Pier*, Benney's *Charity Main* and Sigal's *Weekend in Dinlock*. Dealing with pre-war South Wales, Lancashire in the 1930s, the North East in 1944 and South Yorkshire in the fifties, all offer a good picture of the colliery village but if one wants the ethos of the miner's life then this is contained in a song which is still a favourite in working-men's clubs and folk-song societies in County Durham. In 1963 Novia Scotia (Harraton) Colliery closed, its men were transferred to the Nottingham coalfield and one of its deputies, George Purdom, celebrated the closure by writing a song [**doc. 6**]. As a folk ballad this is an interesting

The Consequences

piece for in just over twenty lines it speaks of hope, frustration, pride and thus becomes part of a tradition which stretches back into the eighteenth century. The men are told, 'Leave your cares behind you, your future has been planned and off you go to Nottingham, to Robens' Promised Land', but not to forget the dead pit or its history for ''Cotia was a colliery, her men were true and bold'. The men recognise that they must change their union banners and endure hearing their children speak another dialect because coal cutting does not bring the security which other workers take for granted. It is a proud, tragic, yet in some ways good-humoured communal song and as such seems to symbolise the traditions of a body of workers who have endured hardship within an industry which has in three hundred years passed through a complex evolutionary cycle and is now in a period of economic decline.

Part Four

DOCUMENTS

document 1

A Coal Canal

Samual Smiles' description of the Worsley Basin gives insight into the enterprise of the first Duke of Bridgewater and his engineer James Brindley. Detailed information about other coal-carrying canals is given in J. Priestley's great work A Historical Account of Inland Navigation and Railroads (1831).

It is at Worsley Basin that the canal enters the bottom of the hill by a subterranean channel which extends for a great distance—connecting the different workings of the mine—so that the coals can be readily transported in boats to their place of sale. It lies at the base of a cliff of sandstone, some hundred feet in height, overhung by luxuriant foliage, beyond which is seen the graceful spire of Worsley church. In contrast to this scenic beauty above, lies the almost stagnant pool beneath. The barges laden with coal emerge from the mine through the two low, semi-circular arches opening at the base of the rock, such being the entrances to the underground workings.

In Brindley's time, this subterranean canal, hewn out of the rock, was only about a mile in length, but it now extends to nearly forty miles in all directions underground. Where the tunnel passed through earth or coal, the arching was of brickwork; but where it passed through rock, it was simply hewn out. This tunnel acts not only as a drain and water-feeder for the canal itself, but as a means of carrying the facilities of the navigation through the very heart of the collieries; and it will readily be seen of how great a value it must have proved in the economical working of the navigation, as well as of the mines, so far as the traffic in coals was concerned.

At every point Brindley's originality and skill were at work. He invented the cranes for the purpose of more readily loading the boats with the boxes filled with the Duke's 'black diamonds'. He also contrived and laid down within the mines a system of underground railways, all leading from the face of the coal, where the miners worked, to the wells which he had made at different points in the tunnels, through which the coals were shot into the boats waiting below to receive them. At

Manchester, where they were unloaded for sale, the contrivances which he employed were equally ingenious.

document 2

John Curr's The Coal Viewer's and Engine Builder's Practical Companion *(1797) is an early engineering textbook. In chapters dealing with methods of conveying coals underground, construction of corves, cast-iron railroads and pumping engines Curr, in simple language, outlines the advantages of the adopting of improved machinery. He also demonstrates practical experience by giving very detailed instructions to blacksmiths on how to repair engines and by presenting elaborate tables on 'fire engine' materials.*

The prevailing practice, till of late in the working of collieries in the neighbourhood of Newcastle-upon-Tyne and Sunderland, was to draw a single corf* on a sled from the workings to the shaft of the pit, which as these workings were extended, and the prices and maintenance of horses enormously increased, became an intolerable burthen to the proprietors of such works; therefore the viewers or superintendents of collieries, have with a great deal of propriety introduced wooden rails, or waggon ways underground, for that purpose, (or what is generally distinguished by the name of Newcastle-roads,) and fixed a frame upon wheels capable of receiving two or three of their basket corves, which upon these carriages and roads are drawn by one horse. But the basket or twig corf which has some great perfections to recommend it at Newcastle-upon-Tyne and Sunderland, where the coals are small, (being of a globular form, with a small aperture at the top,) cannot with propriety be introduced in the southern parts of this kingdom, where the coals delivered to market are all, or in a great measure, large.

And notwithstanding this great improvement, I am of opinion that a greater acquisition is still to be made with the same corf, by laying cast iron roads and placing the corf upon a small frame or tram made upon a proper principle, and hooking or chaining one tram to another.

* A machine made of wood or twigs, in which the coals are drawn from the face of the vein or bed, to the bottom of, and up the shafts.

Having for the above mentioned reasons introduced machines for drawing coals at two of His Grace the Duke of Norfolk's Collieries, near Sheffield, I had still a difficult point to accomplish, which was, to contrive an easy and expeditious mode of conveying the coals to the bottom of the pit, in which I have been successful, far beyond my expectations, and perhaps have hit upon a mode superior to any thing heretofore practised, as the result of seven years experience informs me; I have therefore herein offered to the public the plans and directions for executing both the roads and corves, and every thing relating to the invention, by which means a horse takes at a moderate draught, nine or ten corves of equal size to those at Newcastle-upon-Tyne and Sunderland, of which, even by their improved mode of conveying, the horse takes only two or three.

document 3

The Employment of Children in Mines

The Report of Children's Employment Commission is rightly seen as a landmark in social history. Ashley's Mines Act (1842) would not have succeeded had public indignation not been aroused by this comprehensive document. The verbatim reports of child witnesses are well known but the comments of the sub-commissioners deserve more attention because they throw light on the mining techniques of various areas as well as giving facts about hours, wages and the lives of mine workers. This extract is from the conclusions of the report. A fuller extract is given in English Historical Documents, *volume xii (1) and the complete report is now republished as volume 6 of the Irish University Press Series of British Parliamentary Papers.*

Report of the Commissioners on the Labour of Women and Children in Mines (1842)

First Report: Conclusions

We find:—

1. That instances occur in which Children are taken into these mines to work as early as four years of age, sometimes five, and between five and six, not unfrequently between six and seven, and often from seven to eight, while from

eight to nine is the ordinary age at which employment in these mines commences.
2. That a very large proportion of the persons employed in carrying on the work of these mines is under thirteen years of age; and a still larger proportion between thirteen and eighteen.
3. That in several districts female Children begin to work in these mines at the same early ages as the male.
4. That the great body of the Children and Young Persons employed in these mines are of the families of the adult workpeople engaged in the pits, or belong to the poorest population in the neighbourhood, and are hired and paid in some districts by the workpeople, but in others by the proprietors or contractors.
5. That there are in some districts also a small number of parish apprentices, who are bound to serve their masters until twenty-one years of age, in an employment in which there is nothing deserving the name of skill to be acquired, under circumstances of frequent ill-treatment, and under the oppressive condition that they shall receive only food and clothing, while their free companions may be obtaining a man's wages.
6. That in many instances much that skill and capital can effect to render the place of work unoppressive, healthy, and safe, is done, often with complete success, as far as regards the healthfulness and comfort of the mines; but that to render them perfectly safe does not appear to be practicable by any means yet known; while in great numbers of instances their condition in regard both to ventilation and drainage is lamentably defective.
7. That the nature of the employment which is assigned to the youngest Children, generally that of 'trapping', requires that they should be in the pit as soon as the work of the day commences, and, according to the present system, that they should not leave the pit before the work of the day is at an end.
11. That, in the districts in which females are taken down into the coal mines, both sexes are employed together in precisely the same kind of labour, and work for the same number of hours; that the girls and boys, and the young men and

young women, and even married women and women with child, commonly work almost naked, and the men, in many mines, quite naked; and that all classes of witnesses bear testimony to the demoralising influence of the employment of females underground.

12. That, in the East of Scotland, a much larger proportion of Children and young Persons are employed in these mines than in other districts, many of whom are girls; and that the chief part of their labour consists in carrying the coals on their backs up steep ladders.

17. That in many cases the Children and Young Persons have little cause of complaint in regard to the treatment they receive from the persons in authority in the mine, or from the colliers; but that in general the younger Children are roughly used by their older companions; while in many mines the conduct of the adult colliers to the Children and Young Persons who assist them is harsh and cruel; the persons in authority in these mines, who must be cognisant of this ill-usage, never interfering to prevent it, and some of them distinctly stating that they do not conceive that they have any right to do so.

24. That in general the Children and Young Persons who work in these mines have sufficient food, and, when above ground, decent and comfortable clothing, their usually high rate of wages securing to them these advantages; but in many cases, more especially in some parts of Yorkshire, in Derbyshire, in South Gloucestershire, and very generally in the East of Scotland, the food is poor in quality, and insufficient in quantity; the Children themselves say that they have not enough to eat; and the Sub-Commissioners describe them as covered with rags, and state that the common excuse they make for confining themselves to their homes on the Sundays, instead of taking recreation in the fresh air, or attending a place of worship, is that they have no clothes to go in; so that in these cases, notwithstanding the intense labour performed by these Children, they do not procure even sufficient food and raiment; in general, however, the Children who are in this unhappy case are the Children of idle and dissolute parents, who spend the hard-earned wages of their offspring at the public house.

We have thus endeavoured to present a faithful account of the 'actual state, condition and treatment' of the Children and Young Persons employed in the 'Collieries and Mines' of the United Kingdom, and 'of the effects of such employment on their Bodily Health': the effects of this employment on their 'Morals', it appears to us, will best be shown by bringing them into view in our next Report, in connexion with the intellectual, moral, and religious state of the whole of that portion of the working population which is included under the terms of our Commission.

All which we humbly certify to Your Majesty.

Thos. Tooke.
T. Southwood Smith.
Leonard Horner.
Robt. J. Saunders.

document 4

Evidence Given Before a Select Committee on Coal

The minutes of the various Select Committees on Coal which met with great regularity between 1829 and 1854 are very profitable sources of information about all aspects of the industry. Famous figures in coal-mining history, like John Buddle and George Stephenson, appeared as witnesses and their lengthy testimonies give clear insight into prevailing working conditions and into attitudes to safety. The index in each report makes reference extremely easy.

3934 Do you consider that any particular law could be made applicable generally to all districts of this country, for the prevention of accidents in coal mines?—I do not think that any general law could be framed which would have that tendency, beyond one or two enactments. I think that a few enactments might be desirable.

3935 Have you come to that in consequence of the different characters of the coal mines in different districts?—I have.

3936 Will you describe the different characters of some of the coal mines which have induced you to come to the conclusion that no such law would be available?—The modes of working coal vary greatly, from the differences in character of the coal seams themselves, and of the coalfield, in which they occur. It is difficult to make any comparison between the immense collieries of the north of England, where the regularity of the strata and the depth of the mines occasion the working of one or two square miles of coal from a single winning and collieries in such districts as South Staffordshire, where the faulty nature of the ground and the shallowness of the seams worked, cause them to be limited on an average to five or ten acres. I am putting the two extreme cases in this instance. The mines on the largest scale in the north of England, and on the smallest scale in South Staffordshire.

3937 Looking at these differences would you be disposed to recommend this Committee to report to The House any particular method of ventilation?—No I think that is a matter which must be left entirely in the hands of the mining engineer.

21 July 1853

Evidence of J. K. Blackwell, Esq., *Minutes of Select Committee on Coal*, 1853.

document 5

A Miner's Bond

This extract contains the major part of two clauses from a binding agreement signed in 1833 between the owners of a colliery and a group of miners. Many such agreements were printed in a standardised form with gaps left in the text so that management could insert specific terms of work. In this instance over a hundred workers made their mark and bound themselves for a year. In the original the document takes up two sides of foolscap and its small print is couched in legal terms. This makes it easy to imagine the truth of the oft-reported accusation that many of the men signed after they had received free beer and did not understand to what they were committing themselves.

Memorandum of Agreement, made the sixteenth day of March in the year of our Lord 1833 between the owners of Seghill colliery of the one part and several other persons whose names or marks are hereunto subscribed of the other part.

The said owners do hereby retain and hire, the said several other parties hereto from the fifth day of April next ensuing, until the fifth day of April which will be in the year 1834, to hew, work, fill, drive and put coals and do such other work as may be necessary for carrying on the said colliery, as they shall be required or directed to do by the said owners . . . at the respective rates and prices, and on the terms, condition, and stipulations, and subject to and under the penalties and forfeitures, hereinafter specified and declared; that is to say, FIRST—The said owners agree to pay the said parties hereby hired, once a fortnight, upon the usual and accustomed day, the wages by them earned, at the following rates, namely, to each hewer for every score of coals wrought out of the whole mine, each score to consist of twenty corves and each corf to contain coals sufficient to fill *solidly* the standard Coal Tub used at the said colliery—or by weight coals sufficient to weigh 6 cwt in the High Main Seam where the seam is 3 ft 10 ins in height and upwards 6s 10d per score and where the seam is 3 ft 6 ins and under 7s 1d score of Round coals handpicked——Should any Hewer be found breaking his Coals unnessarily in either seam to be fined 2s 6d for every offence——Each person for whom the said Owner shall provide a Dwelling House as part of his wages shall be supplied with a reasonable quantity of Fire Coal paying to the said Owner 3d per week for leading the same. THIRD—The said owners——shall provide and keep at each pit a tub to contain 87·249 imperial Gall. which shall be equivalent to a 20 peck Corf or a weighing machine: in which case the coals to be 6 cwt and whenever any corves shall be sent to bank suspected to be deficient in measure or weight the coals therein shall be measured or weighed by the heap keeper, or other person appointed for that purpose by the owners, and if found deficient, no payment shall be made for hewing and filling the same but the hewer hereof shall not be subject to any forfeiture or penalty on that account, neither shall any payment be made for hewing and filling any corf in which shall be found stone or splint to the amount of 1 quart or in Separation working, in case round coal shall be mixt with small.

Names and Marks of 71 Hewers
29 Putters
12 Drivers

Collier Songs

document 6

The first song is included because the folk and popular nineteenth-century music of Tyneside is a valuable source of information about custom and dialect. 'The Collier's Rant', 'Cussie Butterfield' and 'Blaydon Races' are well known but are only a small part of a rich tradition of song which exists and continues to grow in the North-East.

The second song shows that the tradition persists.

1. THE BLACKLEG MINER *Mid-Nineteenth Century*

Oh, early in the evenin' just after dark,
The blackleg miners creep te wark,
Wi' their moleskin trousers an' dorty short,
There go the blackleg miners!

They take their picks an' doon they go
Te dig the coal that lies belaw,
An' there's not a woman in this toon-raw
Will look at a blackleg miner.

Oh, Delaval is a terrible place,
They rub wet clay in a blackleg's face,
An' roond the pit-heaps they run a foot-race
Wi' the dorty blackleg miners.

Now, don't go near the Seghill Mine.
Across the way they stretch a line,
Te catch the throat an' break the spine
O' the dorty blackleg miners.

They'll take your tools an' duds as well,
An' hoy them doon the pit o' hell.
It's doon ye go, an' fare ye well,
Ye dorty blackleg miners!

Se join the union while ye may.
Don't wait till your dyin' day,
For that may not be far away,
Ye dorty blackleg miners!

2.   YE BRAVE BOLD MEN OF 'COTIA   1964

Ye brave bold men of 'Cotia,
The time is drawing near
You'll have to change your language, lads,
You'll have to change your beer.
But leave your picks behind you,
You'll ne'er need them again,
And off you go to Nottingham,
Join Robens' merry men.

Ye brave bold men of 'Cotia,
The time is drawing thus.
You'll have to change your banner, lads,
And join the exodus.
But leave your cares behind you,
Your future has been planned,
And off you go to Nottingham,
To Robens' Promised Land.

Ye brave bold men of 'Cotia,
To you I say farewell.
And somebody will some day
The 'Cotia story tell.
But leave your cares behind you,
The death-knell has been tolled.
'Cotia was a colliery.
Her men were true and bold.

document 7

A Glossary of Mining Terms

A Glossary of Terms used in the Coal Trade of Northumberland and Durham *was first published anonymously in 1849 but the third edition in 1888 showed that the author was G. C. Greenwell, a colliery viewer and a President of the North of England Institute of Mining and Mechanical Engineers. In the preface he states that he intends to record 'how things were 50 years ago, and much more'. The book is valuable, for it defines technical terms and gives details of north-eastern mining community custom.*

Geordy—The safety-lamp invented by George Stephenson.

Goaf—A space from which the coal pillars have been extracted it is usually in the first instance a large dome, resting as the extent increased upon the wreck which has fallen from the roof of the exhausted space. Eventually the pressure to a large extent re-consolidates the whole, the surface subsiding.

Gob-fire—Spontaneous combustion in a goaf. Very rare in the North of England.

Hough—'The posterior part of the knee joint.'—The ankle bones more or less completely united. (Webster's Dictionary.)

'Crook your hough!' the friendly salutation of a pitman who wants you to sit down and 'have a crack' scarcely allows hough to have this meaning. It means either to sit on a seat, or on your hunkers; originally, in all probability, the latter.

Hover—'Hover a bit.'—To pause and consider before action.

'*How!*'—To which the reply is 'How again!' The salutation and response of two pitmen, near to or within hail of each other. It may be friendly or otherwise, but is usually the former.

Howdie—A midwife.

Huddock, Huddick—The cabin of a keel.

Hunkers—Sitting on the hunkers. Sitting with the balls of the feet upon the ground and the knees bent, so that the thighs rest on the calves of the legs. This position no doubt became habitual to pitmen from the nature of their underground work, and the conditions under which it is performed.

Jack—During sinking, whilst the two pits or a pit and a staple are being sunk simultaneously by means of two gins, one of them to prevent mistakes, is usually called a Jack.

Methodism and Mining Trade-unionism
document 8

In the early nineteenth century Methodists not only avoided trade union involvement they openly condemned the infant organisations. In 1833 the Wesleyan Conference advised its members to play no part in 'associations which are subversive of the principles of true and proper liberty, employing unlawful oaths and threats and force to acquire new members'. The extract given below illustrates such attitudes and was directed at miners who stopped work in 1810 because they opposed an alteration in the terms of the yearly binding.

Sunderland, 18th November, 1810

AT A MEETING
OF THE
TRAVELLING AND LOCAL PREACHERS,
HELD THIS DAY
IT WAS RESOLVED TO SEND
The following ADDRESS to the Persons Concerned
IN THE SOCIETIES, IN THIS CIRCUIT

To those Members of the Methodist Societies, who have refused to fulfil their Engagements in the Collieries.

The connection we have with you, as Members of a Religious Society, leads us to consider it a duty we OWE TO YOU, TO THE COUNTRY, AND TO GOD, to express our opinion on this part of your conduct.

We do not interfere with any question between you and your Employers, but confine ourselves solely to this one point:—

Is your refusing to fulfil your Contract pleasing or displeasing to God?

Have you considered the consequences of your present conduct?

Can it be justified to the sober part of mankind?

Are you not acting in direct opposition to the Command of God?

Does He not number 'Covenant Breakers' amongst the worst of men? See Romans, i. 31.

Does He not say a good man 'sweareth to his own hurt, and CHANGETH NOT!' Psalm xv. 4.

Does it not bring a scandal on your Profession?

Does it not bring guilt upon your minds?

Are you not making work for repentance in your last hours?

Will the Judge of quick and dead say 'well done', respecting this matter, at the last day?

You know us—you know we seek your welfare.—We 'watch over you as those who must give an account—We shun not to declare the whole counsel of God—We have not ceased to teach you that there can be no piety towards God, without justice towards man.—Judge then our sorrow to see some of you uniting with those 'who fear not God, nor regard man', uniting with them in setting the laws of your Country at defiance.

We entreat you—as you have any regard for justice or truth,—as you would not render your profession or Religion a reproach—as you would not grieve every good man, and above all, as you would not persist in disobedience to God—return to your duty.

Let not the fear of man deter you,—honestly acknowledge your error to your employers,—solicit their protection from those who would hinder you from discharging your duty, and be determined to suffer rather than sin.

Perhaps some of you have been hindered by Fear—others have been misled by false Reasonings.—If however after we have thus affectionately warned you, and waited for your Repentance, you persist in Evil, we must free ourselves from the reproach of your conduct.

Signed in behalf of the Meeting,

William Atherton

document 9

A Novelist's Impression of a Mining Community

Disraeli's description of a mining community was published in 1845 and illustrates a popular response to the three-year-old Report of Children's Employment Commission. The trapper sitting in the dark and the girl harnessed so that she could hurry coal were entering literature as figures to pity.

Far as the eye could reach, and the region was level except where a range of limestone hills formed its distant limit, a wilderness of cottages, or tenements that were hardly entitled to a higher name, were scattered for many miles over the land; some detached, some collected in little rows, some clustering in groups, yet rarely forming continuous streets, but interspersed with blazing furnaces, heaps of burning coal, and piles of smouldering ironstone; while forges and engine-chimneys roared and puffed in all directions, and indicated the frequent presence of the mouth of the mine, and the bank of the coal-pit. Notwithstanding the whole country might be compared to a vast rabbit-warren, it was nevertheless intersected with canals, crossing each other at various levels; and though the subterranean operations were prosecuted with such avidity that it was not uncommon to observe whole rows of houses awry, from the shifting and hollow nature of the land, still, intermingled with heaps of mineral refuse, or of metallic dross, patches of the surface might here and there be recognised, covered, as if in mockery, with grass and corn, looking very much like those gentlemen's sons that we used to read of in our youth, stolen by the chimney-sweeps, and giving some intimations of their breeding beneath their grimy livery. But a tree or a shrub, such an existence was unknown in this dingy rather than dreary region. They came forth: the mine delivers its gang and the pit its bondsmen; the forge is silent and the engine is still. The plain is covered with the swarming multitude: banks of stalwart men, broad-chested and muscular, wet with toil, and black as the children of the tropics; troops of youth, alas! of both sexes, though neither their raiment nor their language indicates the difference; all are clad in male attire, and oaths that men might shudder at, issue from lips born to breathe words of sweetness. Yet, these are to be, some are, the mothers of England! But can we wonder at the hideous coarseness of their language when we remember the savage rudeness of their lives? Naked to the waist, an iron chain fastened to a belt of leather runs between their legs clad in canvas trousers, while on hands and feet an English girl for twelve, sometimes for sixteen hours a day, hauls and hurries tubs of coals up subterranean roads, dark, precipitous, and plashy; circumstances that seem to have escaped the notice of the Society for the Abolition of Negro Slavery. Those worthy gentlemen, too, appear to have been singularly unconscious of the sufferings of the little trappers, which was remarkable as many of them were in their own employ.

Disraeli, B. *Sybil*, 1845

A Collier's Home   **document 10**

Simonin's description of a collier's home life seems romanticised, but most authorities agree that miners enjoyed a relatively high standard of living. The value of this book, however, lies as much in its high-quality engraving as in the information which it gives about mining practices throughout the world.

It is in this country [England] above all others, that the dwellings of colliers should be visited. The cottages in the best districts are ornamental, neat, in many instances detached, and even wide apart. The wife is at home, attentive and thrifty, making the husband's tea or the traditional pudding, and ready to do all that is required in the household. The furniture shines with a bright polish; order prevails everywhere; articles of luxury are seen, books, and a newspaper, which is so often found to be wanting elsewhere. The English miners have even journals of their own, which the French have never yet possessed. The children, of whom there are frequently a troop, are peaceably disposed, and are carefully dressed. It is not the house of a workman, it is rather that of a citizen; it is the cherished home of the Englishman, the sacred and inviolable domestic hearth, such as has given rise to the proverbial saying—Every Englishman's home is his castle.

Simonin, L., *Underground Life or Mines and Miners*, 1868.

document 11
A Mine Inspector's Report on the Use of Safety Lamps

The Reports of the Mines Inspectors are part of a body of Government statistics which give detailed information about the mining industry. They list the number of workers killed, analyse the causes of death and most important of all, suggest ways to improve safety regulations.

Six persons have been burnt to death by firedamp when using the Davy lamp.

The principle of this lamp, and the judicious employment of it for preventing explosion, are not thoroughly understood by the workmen and overlookers of collieries; and it would be well if they studied the history of this valuable invention, and the directions for making and using it that appear in Sir Humphry Davy's Treatise (published in 1818) on flame and the safety lamp.

There is good ground for believing that fatal accidents have occurred with this lamp through ignorance of its real nature and capabilities.

Davy recommended a wire gauze, containing not fewer than 784 apertures in a square inch; he advised that the framework and fittings of the lamp should be so arranged as to prevent the possibility of there being a larger external aperture in any part of it; he warned the miner not to expose it to a rapid current of inflammable air unless protected by a shield half encircling the gauze; and he deprecated the practice of continuing to work with it after the wire attained a red heat.

In my visits to the collieries I have occasionally observed and pointed out to persons present their practical disregard of the foregoing conditions.

Lamps are now made which contain only 550 or 600 apertures per square inch; the framework is in some instances so loosely put together, or so far dilapidated, as to exhibit even larger openings; shields are seldom met with; and the gauze is seen at times red hot, and smeared outside with grease and coal dust, whereby the risk of ignition is rendered more imminent.

When the Davy lamp is thus illegitimately constructed, imperfectly repaired, and improperly used, it is not surprising that it should fail to be an effectual preservative against explosion.

The prevailing notion among miners is that the lamp is infallible; and the majority of them have yet to learn that under certain conditions, defined by its illustrious inventor, it ceases to be what they have hitherto believed it.

In the collieries of the Midland Counties safety lamps are employed only for purposes of experiment and trial; but in those of Yorkshire, which are more fiery, they are in frequent requisition, and several pits are lighted by them entirely.

The Davy lamp is usually adopted, but it is disliked by the miner on account of the feeble illumination yielded by it; and to obtain a better light he is sometimes tempted to take off the gauze, and by so doing he jeopardises his own and his fellow workmen's lives.

Charles Morton, *First Mine Inspector's Report*, 1851.

A Mine Inspector's Evidence

document 12

Richard Fynes' The Miners of Northumberland and Durham is a book before its time. The Webbs are generally considered to be the pioneers of trade union history but Fynes' book had been published twenty years when the History of Trade Unionism came out in 1893. This extract which is taken from the chapter on the Hartley Disaster (1862) typifies Fynes' technique of giving full quotations and comes from a report of a protest meeting held in Newcastle.

MINER—I believe you have something like 150 collieries to inspect?

MR DUNN—Yes.

MINER—Twenty-eight in Cumberland?

MR DUNN—Yes.

MINER—Do you think you are able to inspect all these?

MR DUNN—Well, the Government thinks I am able, you know.

ANOTHER MINER—Were you satisfied with the one shaft at this colliery, if so there is an end to the matter; if not, what steps did you take to remedy the defect? Did you apply to the Secretary of State, showing him that it was defective?

MR DUNN—At this very moment there are three of the largest collieries in Northumberland—Seaton Delaval, North Seaton, and Newsham—managed by the most talented men in Northumberland, all with single shafts. Now, what would you have me to do? Do you think it is my duty to call in question the management of these pits?

MINER—Am I to understand this is an answer to my question?

MR DUNN—Well, I am not so well satisfied as if they had two, but I have not the power to alter it.

Richard Fynes, *The Miners of Northumberland and Durham*, p. 198.

Colliery Rules
document 13

This abridged set of colliery rules is from 1856 and therefore immediately post-dates the act which made publication of such rules obligatory. The printed pamphlet from which these details were taken was of standardised form and almost identical sets of rulings can be found for adjacent pits. A preface giving the terms of authorisation in the 1855 Mining Act often preceded the statement of the seven general rules. Next came the special rules for the colliery. At the Providence Pit, Wakefield, sixty special rules were authorised and, although some of these touch on other matters, most were concerned with safety regulations. At the end the Mine Inspector's handwritten statement of certification recognises the growing power of these Government officials.

GENERAL RULES

The following General Rules shall be observed by the Owner and Agent of the Colliery:

1. An adequate amount of ventilation shall be constantly produced in the Colliery, to dilute and render harmless noxious gases, to such an extent as that the working places of the pits and levels of such Colliery shall under ordinary circumstances be in a fit state for working.

2. Every shaft or pit which is out of use or used only as an air pit, shall be securely fenced.

3. Every working and pumping pit or shaft, shall be properly fenced when not at work.

4. Every working and pumping pit or shaft, where the natural strata under ordinary circumstances are not safe, shall be securely cased or lined.

5. Every working pit or shaft shall be provided with some proper means of signalling from the bottom of the shaft to the surface, and from the surface to the bottom of the shaft.

6. A proper indicator to shew the position of the load in the pit or shaft, and also an adequate brake shall be attached to every machine worked by steam or water power used for lowering or raising persons.

7. Every steam boiler shall be provided with a proper steam gauge, water gauge, and safety valve.

SPECIAL RULES

14. The miners are to build packwalls, and set a sufficient quantity of props for safely supporting the roof, and to renew them as often as necessary, or when ordered by the under-viewer or his deputy. Any person disobeying this Rule must be reported to the agent.

15. The under-viewer or his deputy must see that the permanent stoppings are well built with stone, or bricks and lime, and plastered on one side, and supported by packing or stowing, and that the doors are hung so that they will fall to of themselves.

16. No person shall try the workings or the goaves for firedamp with a candle, and any person smoking tobacco, or having a naked light or lucifer matches, where safety lamps are ordered to be used, must be reported to the agent or under-viewer.

21. When a man is using a safety lamp, he must see that his hurrier does not use a naked light in the same working place.

24. The doors in the principal air-ways must be checked or doubled, and doors which are only used occasionally by the under-viewer or his deputy must be locked.

27. Boys under ten years of age are not to enter the mine and any person evading this order, or misrepresenting the age, will be fined according to Act of Parliament.

31. The pits are to be ready to draw coal at Six o'clock in the morning.

CERTIFICATION (*hand written*) Wakefield, 3rd May 1861
Thereby certify that the foregoing is a copy of the Special Rules established and in force at Providence Colliery in the year 1856, and from that time continuously to the present day.

(*signed*) Chas. Morton
Inspector of Mines

General and Special Rules, Providence Colliery, Wakefield, 1856.

document 14
Management Examinations

After 1872 colliery managers had to have a certificate of competency and to ensure that they passed the examination books of instruction were published. William Hopton's A Conversation on Mines between a father and son to which is added questions and answers to assist candidates to obtain certificates for the management of collieries *uses the very popular 'question and answer' form and is an immediate response to a new demand.*

EXAMINERS—We presume you to be a steady and sober and attentive person.

APPLICANT—I have not had for years any intoxicating drinks, nor ever been drunk.

EXAMINERS—Can you show any proof of your sobriety and attentiveness?

APPLICANT—I have a note here from my employer, with whom I have served for years, one from the manager of the colliery, and another from our minister, and I have also a temperance certificate.

EXAMINERS—You have had a situation, no doubt in mines?

APPLICANT—I have worked my way up from a fireman to a deputy, and from a deputy to an under-viewer or under-looker.

EXAMINERS—Then you are a thorough practical man?

APPLICANT—My experience in mines began in early life.

EXAMINERS—Have you received a mathematical education?

APPLICANT—A little; but in mathematical knowledge I am self taught.

EXAMINERS—You have some knowledge, we presume, of decimals and fractions?

APPLICANT—Yes. My information of figures will enable me to measure, I think, anything required in the management of mines.

EXAMINERS—You will be glad to give us, no doubt, a proof of your knowledge?

APPLICANT—I will give you a few measurements with which near all things in mines and collieries may be measured, or the contents obtained: such as pit boxes, waggons, air ways, currents of air, timber, coal, circular-measurements or the cubical measurements.

Hopton, W., *A Conversation on Mines between a father and son*, Wigan, 1873.

document 15

An Act of Parliament

In a selection of documents about coal mining it seems important to include an Act of Parliament. This eighteenth-century law is included because it is short and because it represents a class of document which is a fruitful source of social history.

Anno decimo tertio
GEORGII II. REGIS

An Act for further and more effectually preventing the wilful and malicious Destruction of Collieries and Coal Works.

Whereas of late divers evil-disposed Persons possessed of or interested in Collieries, have, by secret and subtil Devices, wilfully and maliciously attempted to drown adjacent Collieries, and have by Means of Water conveyed or obstructed for that Purpose, destroyed or damaged the same, intending thereby to enhance the Price of Coals, and gain the Monopoly thereof: And whereas by an Act made in the Tenth Year of the Reign of his present Majesty it was enacted, That if any Person or Persons shall wilfully and maliciously set on fire, or cause to be set on fire, any Mine, Pit, or Delph of Coal, or Cannel Coal, every Person so offending, being thereof lawfully convicted, shall be adjudged guilty of Felony, and shall suffer Death as in Cases of Felony, without Benefit of Clergy: And whereas it is reasonable that an adequate Punishment should likewise be inflicted on Persons who shall wilfully and maliciously destroy or damage Collieries by Means of Water, as is aforesaid; be it enacted by the King's most Excellent Majesty, by and with the Advice and Consent of the Lords Spiritual and Temporal, and Commons, in this present Parliament assembled, and by the Authority of the same, That if any Person from and after the Twelth Day of June, One thousand seven hundred and forty, shall unlawfully, wilfully and maliciously divert, or cause to be diverted Water from any River, Brook, Watercourse, Channel, or Land flood, or convey, or cause to be conveyed Water into any Coal Work, Mine, Pit, or Delph of Coal, or into any subterraneous Cavities or Passages, or make, or cause to be made any subterraneous Cavities or Passages with Design thereby to destroy or damage any Coal Work, Mine, Pit, or Delph of Coal belonging to any other Person or Persons, or shall for that Purpose unlawfully, wilfully, and maliciously destroy or obstruct any Sough or Sewer (which has been a Sough or Sewer in common for Fifty Years) made for draining any Coal Work, Mine, Pit, or Delph of Coal, or shall attempt or continue any such mischievous Practice, or shall aid or assist therein in Manner aforesaid, every such Person shall, for every such Offence, forfeit and pay to the Party or Parties agrieved Treble Damages, and full Costs of Suit, to be sued for and recovered by Action of Debt, Bill, Plaint, or Information in any of his Majesty's Courts of Record at Westminster.

Provided always, That nothing in this Act contained shall prevent or restrain, or be construed to prevent or restrain any Person or Persons, being the Owner or Owners of any Sough, Drain, or Sewer, from destroying, obstructing, or diverting, using, or disposing of any such Sough, Drain, or Sewer, in such Manner as he, she, or they respectively may now lawfully do.

FINIS

document 16

A Miners' Union Newspaper Report

The Miners' Advocate *first appeared in December 1843 having changed its name from the* Miners' Journal. *In its first editorial, beneath a picture of a miner carrying home his tools framed by the national emblems, it proclaimed its allegiance to a country-wide union and to support of W. P. Roberts. Its mottoes were 'Union is strength. Knowledge is power', and 'Let us live by our Labour'. Unfortunately it did not survive for long. After the 1844 strike it collapsed, yet it shares with Tower's* British Miner *the distinction of being a newspaper which was not only supported by miners but also largely written by them.*

THE MINERS' ADVOCATE   27th January, 1844   Price 1½d.

Union is strength.
Knowledge is power.
Let us live by our Labour.

ANOTHER GLORIOUS TRIUMPH FOR THE MINERS ASSOCIATION

Bail Court Westminster January 13th before Judge Williams—Three Miners committed by the Bilston South Staffordshire Magistrates, for leaving their employ, were brought up habeas corpus by the governor of Stafford Gaol, accompanied by Mr W. P. Roberts, the Miners' Legal Adviser, who engaged Mr Bodkin as counsel for the men: counsel for the masters, Mr V. Lea. After hearing the case full argued on each side, THE LEARNED JUDGE QUASHED THE CONVICTION AND THE MEN WERE IMMEDIATELY SET AT LIBERTY! Huzzah! Where is the souless slave that will not join the union?

The Miners' Advocate, Sunderland, 1844.

document 17

A Parliamentary Committee Report on the Featherstone Disturbance 1893

On one occasion when Asquith addressed a political meeting a voice from the floor yelled, 'Why did you murder the miners in Featherstone in '92?' The politician retorted with a nice regard for accuracy, 'It was not '92, it was '93.' It was an unfortunate answer from a man who had been Home Secretary at the time and appointed the committee which inquired into the Featherstone disturbances.

In July 1893 the M.F.G.B. asked members to resist wage reductions of between 10 per cent and 25 per cent. Most miners outside the North East responded and in Yorkshire alone 80,000 were idle and upwards of 250 pits closed. By the end of August the men were becoming restless and when colliers in Barnsley and Wakefield broke the peace the authorities began to prepare for strong action. Police had been withdrawn from the mining districts to control crowds at Doncaster Races, therefore the magistrates first alerted the York garrison and then called out soldiers to disperse a crowd which had gathered at Featherstone near Pontefract to insist that the local pit manager forbid the dispatch of low-grade stockpiled coal. What started as argument about loading smudge escalated into a disturbance in which waggons were overturned and woodpiles set alight. Eventually the soldiers arrived but after an initial good-humoured reception stone-throwing took place. The Riot Act was then read and an hour later, after a bayonet charge had failed to disperse the crowd, two volleys of shots were fired. The second volley killed two innocent bystanders, injured sixteen other men and therefore after a coroner's jury had brought in a verdict which implied censure of the magistrates the Government, conscious of a public outcry, appointed a commission of inquiry.

It is possible to see in this body's report an attempt to excuse misjudgement and placate public opinion. In the eyes of one commentator 'the blood oozed through the white wash'. Leading questions about the central government's direction to the military to hold themselves in readiness were never put and working-men representatives were not appointed Commissioners. On the other hand it should be said that a wide cross-section of witnesses gave evidence and their testimonies in general confirm that a section of the crowd stoned the troops for a long time before the order to fire was given.

The taking of life can only be justified by the necessity for protecting persons or property against various forms of violent crime, or by the necessity of dispersing a riotous crowd which is dangerous unless dispersed, or in the case of persons whose conduct has become felonious through disobedience to the provisions of the Riot Act, and who resist the attempt to disperse or apprehend them. The riotous crowd at the Ackton Hall Colliery was one whose danger consisted in its manifest design violently to set fire and do serious damage to the colliery property, and in pursuit of that object to assault those upon the colliery premises. It was a crowd accordingly which threatened serious outrage, amounting to felony, to property and persons, and it became the duty of all peaceable subjects to assist in preventing this. The necessary prevention of such outrage on person and property justifies the guardians of the peace in the employment against a riotous crowd of even deadly weapons.

Bibliography

Robert Galloway's books on coal mining, although written seventy years ago, have never been superseded. They provide an excellent introduction to the subject by giving detail on most aspects of the industry as it existed before 1850. The shorter of the two books has recently been reprinted and includes a very full modern bibliography by Baron Duckham.

1. Galloway, R. L., *Annals of Coal Mining and the Coal Trade*, Colliery Guardian Co. 1898.
2. Galloway, R. L., *A History of Coal Mining in Great Britain*, David & Charles Reprints 1969.
3. *Report of the Commissioners on Coal 1871*, Report of Committee E.
4. Blake, J. B., 'Medieval Coal Trade of the North East', *Northern History*, vol. 2, 1967.
5. Stone, L., 'An Elizabethan Coal Mine', *Econ. Hist. Rev.*, 2nd series, iii, 1950–1.
6. Nicholls, H. G., *The Forest of Dean* (1858). New edition, David & Charles 1966.

Nef's important work deals with the industry before 1700 and contains extensive tabulated detail on the development of the coal trade.

7. Nef, J. U., *The Rise of the British Coal Industry*, 2 vols, 1932. Reprint, Cass 1969.
8. Sweezy, P. M., *Monopoly & Competition in the English Coal Trade, 1550–1850*, Harvard U.P. 1938.
9. Darby, H. C., 'The Clearing of the English Woodlands', *Geography*, xxxvi, 1951.

Frank Graham of Newcastle has recently produced two facsimile books which deal with coal mining. *The Compleat Collier* is a short and, at 6s, cheap account of early mining techniques and T. H. Hair's *Sketches of the Coal Mines in Northumberland and Durham* is a reprint of a book of lithographs of colliery views.

10. J.C., *The Compleat Collier*, Newcastle (1708). Reprint, Frank Graham 1968.
11. Jenkins, W. J., 'The Early History of Coal Mining in the Black Country', *Trans. Newcomen Soc.*, viii, 1927–8.

12 *Historical Review of Coal Mining*, Mining Assoc. of Great Britain 1924.
13 Needham, J., 'The Pre-Natal History of the Steam Engine', *Trans. Newcomen Soc.*, xxxv, 1962–3.
14 Smith, R., *Sea Coal for London*, Longmans 1961.
15 Mott, R. A., 'Dud Dudley and the Early Coal-Iron Industry', *Trans., Newcomen Soc.* xv, 1934–5.
16 Mott, R. A., 'The Newcomen Engine in the 18th Century', *Trans., Newcomen Soc.* xxxv, 1962–3.
17 Hall, J. W., 'Notes on Coalbrookdale and the Darbys', *Trans., Newcomen Soc.* v, 1924–5.
18 Raistrick, A., *A Dynasty of Iron Founders*, Longmans 1953.
19 Raistrick, A., 'The Steam Engine on Tyneside' (1715–78), *Trans., Newcomen Soc.* xvii, 1936–7.
20 Gale, W. K. V., *The British Iron and Steel Industry*, David & Charles 1967.
21 Rolt, L. T. C., *James Watt*, Batsford 1962.
22 Rolt, L. T. C., *George & Robert Stephenson*, Longmans 1960.
23 Rolt, L. T. C., *Thomas Newcomen*, David & Charles 1963.
24 Harris, J. R., 'The Employment of Steam Power in the 18th Century', *History*, vol. lii, no. 175, 1967.
25 Clayton, A. K., 'The Newcomen Type Engine at Elsecar', *Trans., Newcomen Soc.* xxxv, 1962–3.
26 Priestley, J., *Historical Account of Inland Navigations and Railroads* (1831). David & Charles Reprints 1968.
27 Hadfield, C., *The Canal Age*, David & Charles 1968.
28 Smith, R. S., 'England's First Rails', *Renaissance & Modern Studies*, vol. iv, Univ. of Nottingham 1960.
29 Baxter, B., *Stone Blocks and Iron Rails*, David & Charles 1966.
30 Lee, C. E., 'Tyneside Tramroads of Northumberland', *Trans., Newcomen Soc.*, xxvi, 1949.

Tyson's Archive Teaching Units provide an excellent introduction to document study. Contained in the two collections listed below are facsimiles of major primary sources of railway and coal-mining history. Details of early waggonways and Stephenson's experiments are included amongst twenty-six documents in *Railways in the Making*. The set of coal-mining papers contains facsimiles of employment, pay and price lists together with several documents which relate to the notorious Felling Colliery.

31 Tyson, J. C. (edit.), 'Railways in the Making', *Archive Teaching Units*, Harold Hill & Son Ltd. 1969.
32 Tyson, J. C. (edit.), 'Coals from Newcastle', *Archive Teaching Units*, Harold Hill & Son Ltd. 1968.
33 Meade, R., *The Coal & Iron Industries of the United Kingdom*, Crosby Lockwood 1882.
34 Gibson, F. A., 'A Compilation of Statistics of the Coal Mining Industry of the United Kingdom', *Western Mail*, Cardiff, 1922.
35 Mitchell, B. R. & Deane, Phyllis, *Abstract of British Statistics*, Cambridge U.P. 1962.

The Irish University Press have recently published 250 'blue books'. The 1833–4 reports on the employment of children in factories are obtainable as are those on the employment of children in mines. Five volumes of Select Committee Reports on Accidents in Coal Mines are also published in this series of British parliamentary papers.

36 *First Report of the Commissioners for inquiring into the employment of conditions of children in Mines* (1842). Reprint, Irish U.P. Series of British Parliamentary Papers 1968.
37 Hammond, J. L. & Barbara, *Lord Shaftesbury*, Pelican 1939.
38 Engels, F., 'Condition of the Working Class in England' (1845). *Marx & Engels on Britain*, Foreign Languages Publishing House, Moscow 1962.
39 Gresley, W. S., *A Glossary of Terms Used in the Coal Trade*, Spon 1883.

Lloyd's *Come All Ye Bold Miners* is an extensive collection of songs and poems about pit life. It is difficult to obtain but fortunately much of the material is included, along with the music, in the final chapter of a paper-back book by the author entitled *Folk Song in England*.

40 Lloyd, A. L., *Folk Song in England*, Panther Arts 1969.
41 Lloyd, A. L., *Come All Ye Bold Miners*, Lawrence & Wishart 1952.
42 Morris, J. H. & William, L. J., *The South-Wales Coal Industry 1841–75*, Cardiff U.P. 1958.
43 Hilton, G. W., 'The Truck Act of 1831', *Econ. Hist. Rev.*, vol. x, no. 3, 1958.
44 Holyoake, G. J., *Self Help by the People. The History of the Rochdale Pioneers*, Swan Sonnenschein, 10th ed., 1893.

45 Hobsbawm, E. J., *Labouring Men*, Weidenfeld & Nicolson 1965.
46 Atkinson, F., *The Great Northern Coalfield 1700–1900*, Barnard Castle 1966.
47 Hair, T. H., *Sketches of the Coal Mines in Northumberland and Durham* (1844). Reprint, Frank Graham, Newcastle 1969.
48 Douglas, D. (edit.), *English Historical Documents*, vol. xii (1 & 2), Unwin 1956.
49 Wearmouth, R., *Methodism and the Trade Unions*, Epworth Press 1959.
50 Burt, T., *From Pitman to Privy Councillor: An Autobiography*, T. Fisher & Unwin Ltd 1924.
51 Hopkinson, G. G., 'The Development of the South Yorkshire & North Derbyshire Coalfield 1500–1775', *Trans. Hunter Arch. Soc.*, vol. vii, 6 (1957).
52 Taylor, A. J., 'Combination in the Mid Nineteenth Century Coal Industry', *Trans. Royal Hist. Soc.*, 5th Series, iii (1953).
53 Taylor, A. J., 'The Sub Contract System in the British Coal Industry', *Studies in the Industrial Revolution*, presented to T. S. Ashton (ed. Pressnell, L. S.), 1960.
54 Taylor, A. J., 'The Third Marquis of Londonderry and the North East Coal Trade', *Durham Univ. Journal*, New Series, vol. xvii (1955).
55 Court, W. H. B., 'A Warwickshire Colliery in the Eighteenth Century', *Econ. Hist. Rev.*, vol. vii, no. 2, 1939.
56 Court, W. H. B., *The Rise of the Midland Industries 1600–1838*, Oxford U.P. 1938.
57 Ashton, T. S. & Sykes, J., *The Coal Industry in the Eighteenth Century*, Manchester University, 1929.
58 Rosen, G., *The History of Miners' Diseases*, Schuman's, New York 1943.
59 *Report of the Select Committee appointed to inquire into the nature . . . of fatal accidents 1835*, Reprint, Irish U.P. 1968.
60 Edmonds, E. L. & O. P., *I Was There—The Memoirs of H. S. Tremenheere*, Shakespear Head Press, Windsor 1965.
61 Hair, P. E. H., 'Mortality from Violence in British Coalfields 1800–1890', *Econ. Hist. Rev.*, vol. xxi, no. 3, 1968.
62 Hair, P. E. H., 'Binding of Pitmen in the North East 1800–09', *Durham Univ. Journal*, 1965.
63 Bryan, A. M., 'H.M. Inspectors of Mines. A Centenary Address', *Trans. Inst. of Mining Engineers*, vol. 109, July 1950.

The Catalogue of Plans of Abandoned Mines helps the industrial archaeologist and geographer, for it gives O.S. Survey references which locate the sites of early coal-mines. Pit names and parish locations are recorded together with details of the minerals worked and the approximate date when a colliery was opened.

64 *The Catalogue of Plans of Abandoned Mines*, H.M.S.O. 1922.

65 Clegg, H. A., Fox, A. & Thompson, A. F., *A History of British Trade Unionism*, Oxford 1964.

66 Webb, S., *History of Trade Unionism*, Longmans 1894.

67 Thompson, A. F., *The Rise of the Working Classes*, Penguin, 1968.

68 Arnot, R. P., *The Miners*, Allen & Unwin 1949.

69 Arnot, R. P., *The Scottish Miners*, Allen & Unwin 1955.

70 Fynes, R., *The Miners of Northumberland & Durham* (1873). Reprint, Summerbell 1923.

71 Evans, E. W., *The Miners of South Wales*, University of Wales 1961.

72 Williams, J. W., *The Derbyshire Miners*, Allen & Unwin 1962.

73 Machin, F., *The Yorkshire Miners*, Nat. Union of Mine Workers, Barnsley 1958.

74 Griffin, A. R., *The Miners of Nottinghamshire*, Allen & Unwin 1962.

75 Challinor, R. & Ripley, B., *The Miners' Association—A Trade Union in the Age of the Chartists*, Lawrence & Wishart 1968.

76 Challinor, R., *Alexander Macdonald and the Miners*, C.P.G.B. History Group Pamphlet 48, 1967.

77 Lewis, E. D., *The Rhondda Valleys*, Phoenix House 1959.

78 McCormick, B. & Williams, J. E., 'The Miners and the Eight Hour Day 1863–1913', *Econ. Hist. Rev.*, 2nd series, xii, 2, 1959.

79 Cole, G. D. H., 'British Trade Unions in the Third Quarter of the Nineteenth Century', *International Review of Social History*, ii, 1937.

Many books have been written about the coal industry in the twentieth century but because that period is not the subject of this essay the bibliography of the final chapter only lists some fictional and sociological studies of mining communities.

80 Coombes, B. L., *The Poor Hands*, Victor Gollancz 1939.

81 Orwell, G., *The Road to Wigan Pier*, Secker & Warburg 1965.

82 Benney, M., *Charity Main*, George Allen & Unwin 1946.

83 Hitchin, G., *Pit Yacker*, Jonathan Cape 1962.

84 Llewellyn, R., *How Green Was My Valley*, Michael Joseph 1939.

85 Chaplin, S., *Thin Seam*, Pergamon Paperback 1967.
A story from this book was adapted by Alan Plater as a musical play.
86 Plater, A., *Close the Coalhouse Door*, Methuen 1969.
A compelling appreciation of the mining industry was made in 1961 when Charles Parker, together with folksingers Ewan MacColl and Peggy Seeger produced a B.B.C. programme which used recordings of miners' voices and songs. A record based on the resultant script is now available.
The Big Hewer, a Radio Ballad by Ewan MacColl and Peggy Seeger. Argo Record Company, RG538 mono.
Jack Elliott of Birtley—Songs and stories of a Durham miner. Leader Records, LEA 4001.

Index

Abergavenny Canal, 21
Abrahams, 'Mahon' William, 64
Accidents, 10, 29, 41, 48, 50, 52, 53, 57, 88
Acts of Parliament, 54–65; 1842, 59, 60, 61; 1850, 60; 1855, 63, Doc. 14; 1860, 63, 64; 1872, 65; 1887, 52, 65; 1896, 52; 1925, 89; 1954–62
Age limitations, 63, 64, Doc. 3 and 14
Agriculture, 8, 9, 29, 45
Air coursing, 51
Amalgamated Association of Miners, 76, 79
American Strike, 88
Applied Science, 11
Arbitration, 63, 74, 76, 77
Archaeological Remains, 3, 7
Arling, 54
Ashley Lord, Antony Ashley Cooper, later Lord Shaftesbury, 59, 60, Doc. 3
Ashton, Thomas, 79
Ashton-under-Lyne Meeting, 76
Asquith, H. H., 116

Barnsley, 84
Barnsley Canal, 21
Barnsley coalfield, 19, 21, 46, 64
Baxter, Bertram, 22
Beaumont, Huntingdon, 21, 24
Bell pits, 6
Berwick, siege of, 4
Bessemer, Sir Henry, 16
Binding, 54, 55, 68, 72, Doc. 5 and 9
Birmingham, 20, 21, 81
Birmingham Conference (1889), 81
Black damp, 6
Blackleg miners, 36, 71, 72, 77, Doc. 6
Blenkinsop, John, 23
Boards of Arbitration and Conciliation, 74

Bond, 55, 68, 70, 72, Doc. 5
Boulton, Matthew, 17, 18
Bounties, 45, 46, 54, Doc. 5
Boyle, Edward, 11
Brandling, Charles, 23, 40
Brecknock Canal, 21
Bressley, William, 70
Brewing, 9, 16
Brick making, 9, 17, 25
Bridgewater, Duke of, 20, Doc. 1
Bridgewater Canal, 20, 21, 23, Doc. 1
Brigg's West Riding Colliery, 33
Bristol coalfield, 16, 35
Bristol conference, (1873), 77
British Miner, The, 75, 115
Broseley, Shropshire, 21
Brown, William, 36
Brunton, William (Ventilator), 47, 51
Buddle, John, 47, 51, 56, 59, Doc. 4
Bunkers (coaling stations), 26
Burt, Thomas, 36, 65, 73, 80, 81, 82
Butty system, 40, 43, 55, 59, 69

Calais, 27
Canal Acts, effect of coal mining on, 20
Canal carriage, 19, 20, 21, 22, 25, Doc. 1
Cardiff, 21
Casson, William, 22
Chaldron, 7
Charcoal burning, 8, 9
'Charity Main', Benney, 89
Charles II, petition to, 67
Chartism, 35, 57, 66, 69
Checkweighman, 44, 64, 73, 74
Chesterfield Canal, 20
Chesterfield Coalfield, 19
Chester-le-Street, County Durham, 38
Children's Employment Commission Report (1842), 28, 37, 38

125

Index

Children in coalmines, 28, 29, 32, 51, 56, 58, 59, 60, 62, 63, 68, 72, Doc. 3 and 9
Choke damp, 5, 6
Clanny, Dr, 47, 49
Clean-air legislation, 5
Close the Coalhouse Door, musical play, 89
Cole, G. D. H., 73
Coalbrookdale, 16, 17
Coal dust explosions, 52
Coal prices, 80
Coal production, 3, 4, 5, 16, 17, 18, 19, 25, 87, 88
Coal sizes, 44
Cobbett, William, 29
Coke production from coal, 9, 10, 16
Colliery Journal, 52
Colliery schools, 37–9
Colliery size, 5, 39, 87
Combination Acts, 67
Commissioners for mines, 35, 37, 38
'Common' Engine, 17–19
Communities, coalmining, 29, 31, 32, 66, 68, 89
Compensation, 78
Conditions underground, 6, 28, 68, 69
Conservativism, of coalminers, 5, 29–31, 66
Continental Conferences, 82
Co-operative collieries, 33
Co-operative stores, 33
Corf, or corves, 43, 64, 66, 72, Docs. 2 and 5
Cornish engine, 17
Cort, Henry, 16
County unions, 73, 74, 75, 83
Cowey, Edward, 30, 80
Crawshay family, 39, 40
Cromford Canal, 20
Cumberland mines, 10
Curr, John, 94

Daniels, William, 70
Darby, Abraham, 15
Darby, dynasty of ironmasters, 15 16
Davy, Humphry, 47, 49, 56, Doc. 11
Dearne and Dove Canal (1793), 19
Deforestation, 8
Depth of mines, 6, 10, 18, 19, 41, 50
Derbyshire coalfield, 20, 59, 71

Difficulties in using coal in industry, 9
Disasters in mines, 10, 35, 41, 46, 48, 50, 52, 53, 57, 65, 88; Penygraig Colliery (1800), 52; Felling Colliery (1812), 48; St. Hilda's Colliery (1839), 57; Risca Colliery (1860), 48; Hartley Colliery (1862), 65, Doc. 12; Oaks Colliery (1866), 46, 48; Senyhenydd (1913), 88
Disraeli, Benjamin, 105
District unions, 68, 69, 75, 78
Doherty's National Association of Trade Unions, 68
Don, river, 19
Douglas, river, 20
Domesday Survey, 3
Drainage, 6, 10, 19
Drift Mines, 6
Dudley, ix, 15, 42, 77
Dudley, Dud, 15
Durham coalfield, 7, 8, 10, 19, 21, 27, 29, 34, 38, 45, 48, 70, 72, 78, 81, 83, 89

Ebbw Valley, 38
Education, 36, 57, 75
Edwards, Enoch, 81
Eight Hours Movement, 6, 78, 80
Electricity, 41
Electricity industry, 88
Elizabeth I, 7, 8
Elsecar colliery engine, 19
Engels, Friedrich, 29, 44
Eviction, 38, 68, 71
Exports, 4, 5, 7, 8, 25, 26, 27, 28, 87, 88
Explosions, 51, 57, 65

Factory workers, 29, 68
Family working units, 7
Featherstone disturbance (1893), 116
Federation of county unions, 80
Felling Colliery disaster (1812), 48
Fenton family, 40
Fenwick, Charles, 36, 82
Fifteenth-century Mining, 4, 6
Fines, 44, 45, 46, 61, Doc. 5
Fire-damp, 6, 7, 10, 11, 51
Fitzwilliam, Earl, 19, 46
Flint mill, 49

Index

Flockton Colliery School, 37
Folklore, 29, 31
Fourness, William (Rotary Air Drum), 51
Fourteenth-century Mining, 4, 27, 39
Free Miners of the Forest of Dean, 4, 7, 67
French mining legislation, 62, 83
Friendly Associated Coal Miners within the township of Wakefield, 67
Friendly Societies, 67, 69
Fynes, Richard, 33, Doc. 12

Galloway, R. L., 42
Galloway, William, 52
Gas Problem in Mines, 5, 6, 10, 47, 49, 50
Gasworks, 26, 88
'General' Rules, 56, Doc. 14
Glamorganshire Canal, 21
Glass manufacturing, 8, 17
Gloucester, 67
Goaf, 43, Doc. 7
Grand National Consolidated Trade Union, 68, 69
Group payment systems, 6

Hair, P. E. H., 47, 50
Halliday, Thomas, 77, 78
Hardie, James Keir, 79, 83
Hartley Colliery disaster (1862), 65, Doc. 12
Hepburn, Thomas, 68, 71
Hetton, County Durham, 24, 35
Hobsbawm, E. J., 33
Holidays, 5
Holmes, J. H. H., 49
Holmes, John, 73
Holyoake, George J., 33
Hopton, William, 112
Hours, 45, 56, 59, Doc. 3
Housing, 38, 68, 71, Doc. 9 and 10

Illness, Coalminers', 29, 46, 47
Ingleton coalfield, 24
Inspection of mines, 46, 47, 51, 53, 58, 60, 61, 78, Doc. 11
International Miners Conference, 82

Irish coalfield, 20, 69
Iron industry, 8, 10, 15, 16, 21, 22, 43, 88
Irwell, river, 20

James I, 10
James, Janet, 32
Joint Sliding Scale Association, 84
Jolimont, Belgium conference, 82
Jude, Martin, 70, 73
Justices of the Peace, 33, 70

Killingworth Colliery, 24, 49
Kippax, Yorkshire, 73

Labour Party, 84, 87
Lambton, Lord, 39, 68
Lancashire and Cheshire Federation, 79
Lancashire Coalfield, 3, 20, 21, 24, 57, 70, 76, 79, 81, 89
Language, ix, 30, 67, 90, Doc. 6 and 7
Leicestershire coalfield, 81
Liberal Party, 34, 84
Lodge, miners' 33, 64, 73, 75
Londonderry, Lord, 28, 38, 59, 71
London trade, 5, 25, 78
Leases, 39
Leeds, 67, 77
Leeds Conference of National Miners' Union (1863), 76
Leeds Conference of National Miners' Union (1873), 73, 77
Life expectancy, 47
Lighting in coalmines, 41, 49, 50
Lime burning, 9
Literacy, 35, 37, Doc. 10
'Longwall' system of working coal, 42

Macdonald, Alexander, 63, 72, 75, 79, 81
Management, 36, 45
Manchester, 21
Mansfield, Lord Justice, 54
Markham, Charles, 88
Martin, John (Air Lock and Fan), 51
Marxist Conference (Paris 1889), 82
Marx, Karl, 75
Masters' associations, 25, 41, 70, 78
Mechanics Institutes, 37

127

Index

Mechanisation, 88
Members of Parliament, miners who were, 33, 34, 65
Mersey, river, 20
Merthyr Tydfil, 21
Metal imports, 16
Metal production, 8, 9, 43
Methodism, 33–5, Doc. 8
Midland coalfield, ix, 3, 18, 25, 48, 77, 79
Middleton Rack Railway, 1812, 23
Militancy, 76, 79, 84
Mineral ownership, 39
Miner's Advocate, 70, Doc. 16
Miners' Association of Great Britain and Northern Ireland, 61, 69, 72, 73, 75
Miners' Federation of Great Britain, 78, 81, 82, 83, 87, Doc. 17
Mines Inspectors, 46, 47, 51, 53, 58, 60, 61
Miners' Journal, 115
Miners' Philanthropic Society, 69
Miners' National Union, 79–82
Mines, Neolithic flint, 6
The Miners' Next Step, pamphlet, 87
Mines Rescue Stations, 41
Minimum wage, 79, 80
Mining Acts, *see* Acts of Parliament
Moderation, 74, 75, 76
Monasteries, mines attached to, 4
Monmouthshire canal, 21
Murray, Matthew, 24
Music tradition, 22, 29, 30, 89, Doc. 6

National Coal Board, 30, 89
Nationalisation, 89
Nationalisation Bill (1946), 89
National Miners' Union, 76, 78
Neath Canal, 22
Neath Valley Coalfield, 21
Newcastle roads (tramways), 22, 23, Doc. 2
Newcastle-upon-Tyne, 4, 5, 7, 10, 17, 26, 33, 67, 70, 72, 78, Doc. 2
Newcastle Yeomanry, 66
Newcastle, Earl of, 39
Newcomen, Thomas, 11, 17, 18, 19
Newport, Monmouthshire conference (1889), 82

Newry Navigation, Northern Ireland' 20
Newton, Isaac, 11
Norfolk, Duke of, Doc. 2
Normansall, John, 44, 64
Northern Star, 69
Northumberland coalfield, 3, 8, 10, 18, 21, 22, 23, 24, 25, 26, 27, 34, 42, 45, 48, 57, 61, 69, 70, 72, 78, 81, 83, 89
Northumberland and Durham Mutual Confident Association, 74
North Wales coalfield, 16, 66, 68, 71
Nottingham coalfield, 42, 71, 80, 89, Doc. 6
Novels about mining, 89
Nova Scotia mine, County Durham, 89
Numbers of workers in twentieth-century mines, 89

Oaks Colliery disaster (1866), 46, 48
Output, 7, 9, 10, 18, 27, 50, 72, 80, 87
Over production, 6, 78
Owen, John, 32
Owen, Robert, 68
Ownership, 39, 87

Paine, Tom, 68
Parliamentary intervention, 54–65, 68, 77, 80, 81, 88, 89
Parliament, miners in, 87
Payment, methods of, 43, 46, 72, Doc. 5
Peaceful demonstrations, 71
Penygraig Colliery disaster (1800), 52
Peterloo Massacre, 67
Pickard, Ben, 64, 73, 79, 82, 84
Pit closure, Doc. 6
Plot, Dr, 9
Poetry, 22, 29, 30, 70
Politics, 66
Ponies, 70
Poor Law Commissioners, 38
Ports, 26, 27
Possibilist conference (Paris 1889), 82
Pillar robbing, 42
'Pillar and Stall' system of working coal, 42
Pitmen Poets, 30–1
Plans of mines, 58
Population explosion, 16

Index

Practical Miners' Association, 76, 77
Price of coal, 7, 81
Priestley, J., 93
Primitive Methodists, 33-6
Production limitation, 78
Profit sharing, 33
Pump, Chinese chain, 10
Pumping engine, 17, 65
Punch, 71
Punishment, 55, 63
Purdom, George, 89

Radicalism, 34, 66, 67, 68
Railways, 8, 16, 21, 22, 24, 32, 41, 78
Railway Company purchases, 25
Railway towns, 25
Regional variations, ix, 29, 48
Religion, 33-7, 54, Doc. 3 and 8
Risca Colliery disaster, South Wales (1860), 48
Reynolds, Richard, 16
Rhondda, 26
Rhuddlan, North Wales riot, 66
Richardson, Edward, 73
Road to Wigan Pier, Orwell, 89
Robens, Alfred, 90, Doc. 6
Roberts, W. P., 32, 70, 76, Doc. 16
Roman, coalmining, 3
Roof falls, 6, 42
Rothwell, Yorkshire, 46, 49
Royal Commission (1924-5), 88
Royal Society, 11
Royalty payments, 39-40

Safety cages, 52
Safety lamps, 47, 49, 53, 56, 65, Doc. 14
St Hilda's Colliery disaster (1839), 57
Samuel, Herbert, 89
Sankey Brook Canal, 20
Savery, Thomas, 11
Seghill Colliery, Doc. 6
Schools, 36-8
Scottish Miners Emancipation Act (1774), 54, 55
Seam sizes, 15, 43, 42
Sea transport, 4-5, 78
Select Committee on Accidents in Mines (1835), 57

Senghenydd disaster (1913), 88
Seventeenth century, 8, 9, 11, 15, 27
Sexual promiscuity, 58, Doc. 3
Selling price of coal, 23
Scottish Coalfield, 3, 10, 18, 31, 42, 43, 45, 54, 59, 68, 75, 76, 81, Doc. 3
Shaft Coal mines, 6, 7, 41, 65
Shaftesbury, Lord (Lord Ashley), 59, 60, Doc. 3
Sheffield Coal mine, 5, 19
Sheffield Trade Unionists, 71
Shift work, 45, 59
Shrewsbury, Earl of, 5
Shropshire coalfields, 3, 15, 21, 43
Simonin, L., 107
Sixteenth-century mining, 6, 7, 8, 9, 10
Size of collieries, 39-41, 48, 50, 88
Slavery, 54
Sliding scale, 78-9, 84
'Sloop and room' method of winning coal, 42
Smeaton, John, 11
Smelting, 9, 15
Smiles, Samuel, 17, 93
Smillie, Robert, 87
Smith, Adam, 45, 55
Smith, Southwood, 58
Socialism, 35, 68, 84
Society of Hoastmen, 7
Somerset Coal Canal, 21
Somerset Coalfield, 21, 81
Song, 29, 70, 90, Doc. 6
South Shields Committee, 57-8
South Wales coalfield, ix, 3, 21, 22, 25, 26, 31, 34, 38, 39, 42, 47, 52, 57, 61, 64, 68, 70, 71, 76, 83, 89
South Yorkshire Miners' Association, 74
'Special' Rules, 64, Doc. 14
Spedding, Caslisle, 47
Spedding, James, 51
Staffordshire coalfield, ix, 15, 31, 32, 33, 34, 68, 76, 79, Doc. 16
Standard of living, 28, 33, 45, 46, 66, 68
Stationary steam engines, 23
Steam coal, 26
Steam driven machinery, 24, 25, 30
Stephenson, George, 15, 24, 49, 65
Stone dusting, 52

129

Index

Strikes, 30, 31, 66–84; Northumberland Strike (1810), 69; Hepburn's Union Strike (1830), 68; Miners' Association Strike (1844), 30, 36, 44, 69, 70–72, Doc. 16; South Staffordshire Strike (1869), 76; Amalgamated Association Lockout (1875), 77; M.F.G.B. Strike/Lockout (1893), 84; Yorkshire Strike (1969), 89
Strike funds, 69, 81
Strike levy, 71, 79, 81
Sun-contracting, 40, 69
Subscriptions, 67, 74
Substitution of coal for wood, 9
Sunderland, 24, 29, 68, Doc. 2, 8 and 16
Sunderland Society for Preventing Accidents in Coal Mines, 49, 56
Swallow, David, 32, 70
Swansea Canal, 21, 22

Temperance Societies, 33, 37, 57
Thirteenth-century Mining, 4, 6
Tied housing, 38, 68
Timber, demand for, 8
'Tommy Shops', 31, 68
Tonypandy, 88
Tooke, Thomas, 58
Tools, 45
Tower, John, 75, Doc. 16
Town Moor Meeting, Newcastle (1844), 72
Toyne, J., 82
Trade unions, 63, 64, 65, 66–84
Tramways, 22, Doc. 1 and 2
Trapper, 51, 58, Doc. 3
Tremenheere, Seymour, 35, 45, 57, 61, 62
Trevithick, Richard, 23
Troops used against miners, 66, 67, 73
Truck Act (1817), 55
Truck system, 29, 31–2, 55, 57, 60, 68
Tufnell, A. Carleton, 57

These Poor Hands, novel, 89
Transport system, 3–5, 7, 18, 19, 20, 21, 22, 23, 24, 25, 32, 41
Triple Alliance, 87
'Twelve Apostles', 73
Tyne, river, 4, 7, 8, 22, Doc. 6
Tyrone coalfield, Ireland, 20

Ventilation problems, 6, 42, 47, 50, 65, 67, Doc. 4 and 14

Wages, 6, 32, 43, 45, 46, 55, 57, 60, 64, 68, 69, 70, 72, 77, 78, 79, 80, 84, 89, Doc. 3
Waggon ways, 22
Wakefield, 19, 37, 67, 69
Warwickshire Coalfield, 5
Water problem in mines, 5, 6, 7
Watt, James, 17, 18
Wearmouth, Robert, 34
Wear River, 8, 22
Webb, Sidney and Beatrice, 73, Doc. 12
Wednesbury Canal, 20
Weekend in Dinlock, Sigal, 89
Welsh Canal, 21
Wesley, John, 34
West Bromwich, 77
West Yorkshire Miners' Association, 79
Whitehaven, William Pit, 17
Wilkinson, John, 17
Willoughby, Francis, 9, 21
Wilson, John, 36, 82
Wiltshire coalfield, 21
Women in coalmines, 28, 56, 58, 59, Doc. 3
Work contracts, 5, 29, 55, 68, 70
Worsley mines, 20, 21

Yorkshire coalfield, ix, 3, 5, 6, 19, 21, 23, 24, 27, 33, 34, 36, 38, 40, 42, 45, 46, 60, 64, 67, 68, 69, 71, 73, 74, 76, 78, 79, 89, Doc. 3
Yorkshire Miners' Association, 73, 79
Young, Arthur, 38